Preface

The Chrysler, Ford, and General Motors tooling and equipment representatives and the Riviera Die and Tool companies have developed a new quality system requirement for the tooling and equipment suppliers. The requirement is called *Quality System Requirements: Tooling and Equipment Supplement* (the TE supplement). The supplement is controlled by the Automotive Task Force and is distributed by the Automotive Industry Action Group (AIAG).

The document is designed to equip all nonproduction tooling and equipment suppliers with the tools they will need to ensure quality in their processes. As such, it applies to all suppliers that produce tooling and equipment used to manufacture cars and trucks (i.e., plating, robotics, assembly). It replaces Chrysler's *TESQA* and Ford's *Facilities and Tooling Quality System Survey*.

Its purpose is to provide

an applicable interpretation of the QS-9000 requirements

common quality system requirements

focus in the application of process-driven quality systems

promotion for the effective use of the reliability and maintainability principles

This book addresses the evolution and the development of the TE Supplement, its rationale, its structure, the interpretation of each of the elements, and its relationship to both ISO 9000 and QS-9000. It also discusses the issues of implementation, documentation, auditing, and certification. It builds on my last book, *Integrating QS-9000 with Your Automotive Quality System* (Stamatis, 1996).

The book is written with two objectives in mind. The first is to provide a reference for the TE Supplement. As a result, the target audience is anyone who is interested in and/or involved with quality issues in the tooling and equipment industry. The second objective is to provide the rationale and applicable examples of appropriate documentation, as well as specific implementation steps, for an effective quality system within the tooling and equipment industry. The target audience here is the individual(s) in any organization responsible for the implementation process.

HOW TO USE THIS BOOK

This book is designed to give readers a complete overview of the TE Supplement. Toward that end, I have provided a cursory background of the historical perspective of automotive quality, as well as an overview of both ISO 9000 and QS-9000.

I have also provided an appendix containing a complete quality manual and an extensive procedure/instructions appendix. In both cases, the reader is encouraged to use them as a model for their own processes. Finally, I hope this book will help readers get a better understanding of what is required to implement the TE Supplement in their organization.

LANCHESTER LIBRARY, Coventry University

Much Park Street, Coventry CVI 2HF Telephone 024 7688 8292

This book is due to be returned not later than the date and
time stamped above. Fines are charged on overdue books

D.H. STAMATIS

IMPLEMENTING THE
TE SUPPLEMENT
TO QS-9000

The Tooling and Equipment
Supplier's Handbook

QUALITY RESOURCES.
A Division of The Kraus Organization Limited
New York, New York

Most Quality Resources books are available at quantity discounts when purchased in bulk. For more information contact:

Special Sales Department
Quality Resources
A Division of the Kraus Organization Limited
902 Broadway   800-247-8519
New York, New York 10010   212-979-8600
www.qualityresources.com   E-mail: info@qualityresources.com

Printed in the United States of America

02 01 00 99 98   10 9 8 7 6 5 4 3 2 1
                        ∞
The paper used in this publication meets the minimum requirements of American National Standard for Information Sciences—Permanence of Paper for Printed Library Materials, ANSI Z39.48-1984.

ISBN 0-527-76335-7

Library of Congress Cataloging-in-Publication Data

Stamatis, D.H., 1947-
        Implementing the TE supplement to QS-9000 : the tooling and equipment supplier's handbook / by D.H. Stamatis.
                p.   cm.
        Includes bibliographical references and index.
        ISBN 0-527-76335-7 (hardcover : alk. paper)
        1. QS-9000 (Standard)—Documentation. 2. Automobile industry and trade—Quality control—Standards—United States. 3. Automobile industry and trade—Quality control—Documentation—United States. 4. Automobiles—Parts—Design and construction—Quality control—Standards—Documentation—United States. I. Title.
TL278.5.S83   1997
629.2'34—dc21   97-26716
                                                           CIP

Figures

Tables

To my unforgettable friend, Nikolao Kozaiti

Contents

Acknowledgments

It is with deepest appreciation that I thank the following persons. Without them, this work would not have been possible.

V. Pasqua from the Paslin Company; J. N. Quinn from BOK Company; R. Buchanan and G. T. Curtis from Pacer, Incorporated; R. Karagitz from Ford Motor Company; H. Kinzey from Hank Kinzey and Associates, Incorporated; and J. E. Lacroix from Bra-Con/X-Mation, Incorporated for their patience over the years in helping me understand the tooling and equipment business.

W. McCarthy, for his thoughtful suggestions throughout the project.

T. D. Stamatis and S. D. Stamatis for their continual support and help with typing and computer work.

E. Rice and R. Munro from Ford Motor Company for their encouragement and, as always, their thoughtful suggestions.

The editors of this book, who, through their persistent help, made the difference in the final product.

In addition, I would like to thank the following for granting me permission to use their material. Specifically, R. Roush, A. Love, and Chrysler for granting me permission to summarize the advanced product planning process from Chrysler's publication, *Product Assurance Planning Manual* (2nd ed., November 1995).

I also thank S. Nickolas and Ford Motor Company for granting me permission to summarize the steps and the status reporting process identified in the *Advanced Product Quality Planning (APQP) Status Reporting Process: Participants Guide* (November 1995). This document is a noncontrolled document of Ford Motor Company, Dearborn, Michigan.

I thank Prentice-Hall, Englewood Cliffs, New Jersey, for granting me permission to use some material on system effectiveness and life cost cycle from the book, *Logistics Engineering and Management* (1986, pages 18–19), written by B. S. Blanchard.

The National Center for Manufacturing Sciences, Incorporated, granted me permission to summarize the five-phase management process from the *Reliability and Maintainability Guideline*, pages 2-2 through 2-4, 4-5, and Table 4-1.

I also thank the Excel Partnership, Incorporated, for granting me permission to use some of their material for the background on ISO. The material is from Chapter 4, Section 1, in the *Lead Auditor Course* (Revision 4.0).

Finally, thank you to my seminar participants, who through their questioning and strong recommendations suggested the writing of this book.

1

An introduction to quality

This chapter gives a cursory review of quality. Specifically, it discusses a definition of quality and factors that influence quality, and addresses the issues of variation and process. It closes with a general historical background of ISO 9000, QS-9000, and the TE Supplement.

OVERVIEW

Throughout human history, quality has contributed significantly to everything we do and how well we do it. Quality applies to both products and services; as of late, it has become the de facto characteristic in world markets. But, what is quality? ISO 8402 defines it as "the totality of features and characteristics of a product or service that bear on its ability to satisfy a stated or implied need." This definition takes into account that users often have requirements that remain unstated. Examples of unstated requirements are the consumer's expectations for user instructions to be supplied with domestic products such as cars, cameras, tools, and so on. Although these expectations are unstated, customer satisfaction will only be achieved if the expectations are met.

The definition also recognizes that quality means achieving the requirements, whatever those requirements may be. The terms *high quality* and *low quality* should not be used when *high grade* or *low grade* are what was intended.

Finally, it must also be recognized that customers' requirements change from time to time. Quality is a moving target. When talking about quality, keep an eye on the target and be able to respond to wherever the target may be.

FACTORS THAT INFLUENCE QUALITY

Even though from a customer's perspective quality quite often is a given characteristic, there are some specific drivers in any quality definition. In fact, these drivers are quite different for products and services. Let us look at the products first and then examine the services.

Products

What we do

We establish existing customer needs. Perhaps one of the most critical factors in defining quality is the issue of the customer and what that customer needs, wants, and expects. Depending on that definition, everything else will follow.

We establish future customer needs that do not exist at present. This is the essence of customer satisfaction and continual improvement. It is expected that you should "delight" the customer with something beyond what is expected.

We establish products that meet the above customer needs and stay ahead of customers and competitors. This is the area in which benchmarking comes to the forefront of quality. It is this assessment that will identify the "best in class" products and accordingly transfer that knowledge in your organization.

We describe the above products in terms of detailed designs and specifications. Unless the design and specifications can be manufactured, it is a futile exercise. It is imperative that the organization know the capability and feasibility of everything it produces. In this way, it helps itself and helps the customer in the process.

We plan and install facilities. To have the best designs and plans is meaningless unless the appropriate facilities exist. Quality quite often is defined by the facilities, as well as the layout of those facilities.

We provide systems for moving ideas and innovations to execution. This is the area in which leadership is essential. Without a strong leader to champion quality, the organization will fall short of expectations. A good quality system—for starters—is the ISO 9000 management system.

How well we do what we do

We do quality work first by executing manufacturing at optimum values with as little variation as possible. The need to control, standardize, and improve process variation must be understood by everyone in the organization. Furthermore, major efforts should be made to understand and reduce variation. To understand variation, one must understand its components: special causes and common causes.

Special causes are temporary or local to the process, or they may come and go sporadically depending on the process action. Evidence of the lack of statistical control is a signal that a special cause is likely to have occurred.

A typical example of a special cause is one secretary making more errors than the other secretaries in filling out a particular form. Investigation shows that this secretary was absent the day training was given. *Action*: Train this secretary and note the importance of this training for all future secretaries. *Moral*: Special causes are controlled by operators and corrected by the intervention of management and/or the operator.

How do you eliminate special causes? By at least three actions:

1. Immediately search for the cause of trouble when control gives a signal that a special cause has occurred. Find out what was different on that occasion from other occasions.
2. Seek ways to prevent that special cause from recurring.
3. Do not make fundamental changes in the process. Instead, seek to eliminate that particular temporary or local problem.

On the other hand, common causes are causes that appear on all occasions and in all places. They are inherent in any process. Even though they are present everywhere, the degree of presence varies. In fact, one may even say that each cause contributes a small effect to the variation in results. The reader should be warned: Variation due to common causes will almost always give results that are in statistical control.

A typical example of a common cause is all secretaries making occasional errors in filling out a particular form. *Action*: Consider changing the form or providing new training to all secretaries. *Moral*: Common causes are controlled and corrected by management.

How do you improve common cause variation? By developing a strategy for improvement. Common causes of variation usually cannot be reduced by trying to explain the difference between high and low points within the statistical control limits. A system subject only to common cause variation is best improved by making fundamental changes to the system. Aspects of the system that are always present will need to be changed to improve common cause systems. All the data, not just high or low points, are relevant to further analyses.

The steps to reduce variation are as follows:

plot the data in time-ordered sequence (run chart)

establish control limits if applicable (control chart)

analyze for special causes

remove the special causes

quantify and predict the system's variation

study the system

work on the system to reduce the system's variation

Second, we reduce complexity during manufacture. In the modern world of quality, the foundation of product quality is the voice of the customer. Setting targets for product attributes will help identify the amount of change required for a new product. To facilitate these attributes and changes in the new product, it is imperative that we also understand the concepts of reusability, commonality, carryover product, and complexity reduction.

Reusability focuses on using facilities and tools from the prior design or job. For component tools, this could include reusing dies, molds, checking fixtures, and so on. For component facilities, this could include reusability of processes, machine tools, or floor space. For assemblies, reusability could include equipment such as conveyors, workstations, transfer equipment, bonding equipment, welding equipment, and so on.

Commonality focuses on product components, assemblies, features, product attributes, facilities, and tools that are shared with other designs, but not necessarily the last design. (The reader should note that if the design shares items from the last design and/or job, by definition it is called *carryover*.) Sharing parts, assemblies, and product attributes will generally reduce the number of new tools and facilities required. Product/process concepts that employ manufacturing flexibility often enable commonality and can also drive reusability on future-generation products.

Carrying over product components, assemblies, or features from the prior design will enable reusability of facilities and tools. Items that have little or no effect on customer-perceived value should be targeted for carryover assumptions.

Complexity focuses on the intricacy of a product. The number of components, assemblies, and available options are examples of complexity. Complexity generally increases the cost of facilities and tools by requiring sophisticated equipment and/or increases the quantity of facilities and tools required because of product uniqueness.

Third, we prevent defects through defect-proofing. The idea of prevention, as opposed to appraising product quality, is emphasized here. It is much easier, as well as more cost effective, to plan for prevention as early as possible rather than inspect the product later in the process. One of the ways to perform defect-proofing is through a Failure Mode and Effect Analysis (FMEA) for both the design and process.

Services

What we do

We establish existing customer needs. Just as in the product category, the customer needs must be understood and satisfied.

We establish future customer needs that do not exist at present. Through continual improvement, customer satisfaction surveys, and so on, it is imperative that future needs, wants, and expectations of customers be addressed and incorporated into services with which the customer will be delighted.

We establish services that meet the above customer needs and stay ahead of customers and competitors. Just as in the case for product, it is imperative that we conduct benchmarking studies to establish the current status, as well as the best in class, of either the competitors or someone else providing the same and/or similar service. Once the best in class has been identified, then efforts must be made to implement them in our organization.

We describe the above services in terms of detailed requirements and specifications. Unless the design and specifications of the service can be delivered, it is a futile exercise. It is imperative that the organization know the capability and feasibility of everything that is served and/or about to be served. In this way, the organization helps itself, but also helps the customer, in the process, not to have false expectations.

We plan and establish the mechanisms and processes for providing the above services. To have the best designs and plans is meaningless unless the appropriate facilities exist and the commitment to deliver the designed service is championed by a leader in the organization. Service quality quite often is defined by the facilities, as well as the layout of those facilities, not to mention the delivery of that service.

How well we do what we do

We execute the services in accordance with the established requirements with as little variation as possible. We must understand the concept of variation and implement actions to reduce it as much as possible (see product discussion above).

We reduce complexity in the processes that are involved in the supply of the services. Again, as in the case of product, we must be cognizant of reusability, commonality, carryover service, and complexity.

We prevent errors by error-proofing the processes. A service, just like a product, can have a process; as such, it must be designed as early as possible with as much as error-proofing as possible.

We develop relations with customers. Perhaps the most important attribute of the service is the characteristic of interpersonal skills in the process of delivering the service. It is imperative that the personnel who handle customer service be well-equipped with excellent communication skills, tact, flexibility, persistence, objectivity, integrity, and, above all, the authority to bring to closure a problem with the customer.

Process

Another important issue in modern quality is the issue of "process." Each of the activities involved in executing the whats and the hows of a task, job, machine, and so on take place by means of a process. This is true for both products and services. Therefore, the provision of a product or a service involves many processes. Each process consists of inputs, a conversion activity, and an output. Of course, behind the inputs are the suppliers; the outputs of the processes go to customers. Furthermore, there is a feedback loop from the customers to the suppliers and a feed-forward loop from the suppliers to the customers. A typical pictorial view of this is shown in Figure 1-1.

Quality addresses all the processes involved in the provision of a product or service with respect to the whats and the hows so these processes can be continuously improved to function better at lower and lower costs.

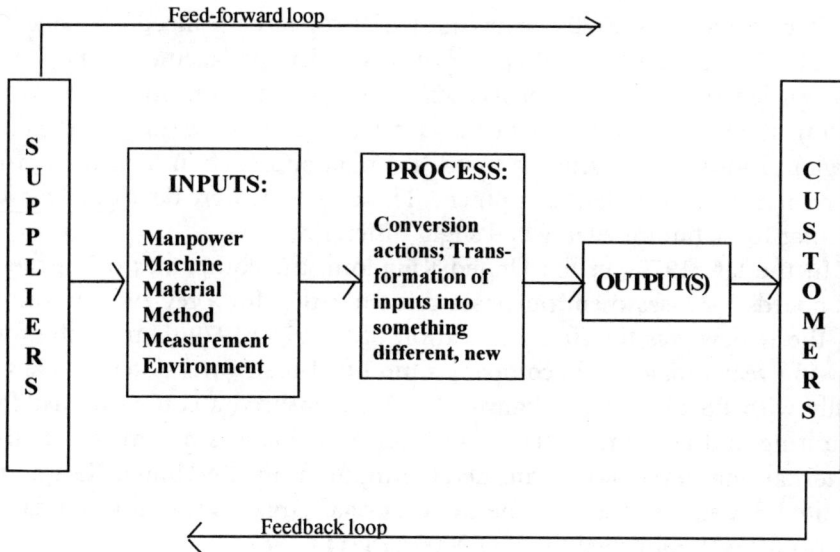

Figure 1-1. A typical process.

HISTORICAL PERSPECTIVE OF ISO 9000, QS-9000, AND THE TE SUPPLEMENT

The international quality system standard series has been a major factor in the European Union (EU) since its adoption in 1987. Historically, however, its development can be traced to quality systems standards developed for the defense industries. In particular, the standards were developed for manufacturing industries; as a result of this history, some terms (such as *product*) reflect its origins (Excel, 1994; Stamatis, 1995b).

The early defense quality systems standards were considered revolutionary in their time. They changed the basis of contracts between purchaser and supplier from mistrust and an emphasis on inspection to supplier responsibility. The early standards also created an awareness that there are many processes and activities in every company that have an effect on the company's ability to satisfy the customer. These processes and activities are in addition to the manufacturing processes.

In the United States, the Department of Defense (DoD) based its purchasing systems on those from its North Atlantic Treaty Organization (NATO) partners; these requirements were developed as MIL-Q-9858 and MIL-I-45208. At the same time, in the United Kingdom, the 50-20 series of military defense purchasing standards were developed.

During the early 1970s, many industrial concerns realized the benefits of adopting a standardized approach to quality management. They also recognized how quality management was distinct from quality control. About the same time, many of the larger manufacturing and purchasing organizations were beginning to use a similar approach in their own procurement of major plant equipment. These systems were developed along similar lines, but inevitably had slight differences.

In the late 1970s in the United Kingdom, the companies using these standards for assessment purposes came together for a review. The result of the review was the first publication of BS 5750:1979. During the following years, major U.K. companies modified their own systems to align fully with BS 5750. This alignment process ensured a common base for auditing and assessment. The significance of this was not missed at the international level. With considerable input from the United Kingdom, United States, and Canada, the International Organization for Standardization (ISO) published the ISO 9000 series in 1987.

By 1992, more than 55 nations had adopted ISO 9000 as a basis for their own equivalent national standards for quality systems. The U.S. standard was adopted as American National Standards Institute/American Society for Quality (ANSI/ASQ—formerly the ASQC) Q90; in the European Union, Standard EN 29000 was published. The international standard is now under publication in over 90 countries.

In 1994, the first revisions to the ISO 9000 series were published. These revisions reflected the phenomenon of third-party registration as the revisions were basically accepted interpretations of some inadequacies of the original 1987 version. The next revision version is expected at some time in the late 1990s.

The international standard is pragmatic in allowing companies to develop systems that are appropriate to their needs and the needs of their customers. A company first determines which markets it is targeting and, as a result, what the company's objectives, are. Once senior management has defined the company objectives they can define the quality policy. Following these decisions by top management, the company can determine how it is going to meet the company objectives, contract requirements, and the requirements of ISO 9000 within the context of the company's culture and technology.

The main emphasis of the standard requires companies to establish what is to be achieved, develop processes designed to prevent errors, and, therefore, place less emphasis on having inspection filters.

The standard highlights the basic management activities that all successful companies need to achieve and is worded to allow freedom for each company to determine the mechanism by which it achieves those requirements. In essence, the standard requires companies to build into their systems methods to measure the extent to which implementation and effectiveness are achieved. The objective of the measurement methods is constant improvement of the system.

By 1987, the standardization of quality systems around the world was here to stay. In the United States, the automotive industry (known as the Big Three) recognized that something had to be done with the plethora of documentation each of the automotive companies was asking the supplier base to produce, document, and keep. As a consequence, in 1988 the purchasing and supply vice presidents of these companies chartered a task force (now known as the Automotive Task Force) to standardize reference manuals, reporting formats, and technical nomenclature. The results were astonishing. Since then, the Automotive Task Force has published five standardized manuals; all of them have been received well by the supplier base.

As the demand for quality was increasing and the ISO 9000 series was becoming ever more popular, in December 1992 the vice presidents of the Big Three authorized the Automotive Task Force to look into the possibility of harmonizing the fundamental supplier quality systems' manuals and assessment tools.

The Automotive Task Force looked at the ISO 9000 series. However, because the standard is a basic quality system, they felt that ISO 9000 did not fulfill the requirements of the automotive industry and that the ISO 9000 standard was not strong enough for their industry. As a consequence, the Automotive Task Force developed their own standard specifically for the U.S. automotive industry and its suppliers, including the following requirements:

ISO 9000 in its entirety, which makes up Section I

additional requirements common to the automotive industry in the United States, known as Section II

additional specific requirements from each of the automotive companies; Section III

This standard was called *Quality System Requirements: QS-9000* (known simply as QS-9000). It was published in August 1994; in February 1995, there was a revision to that standard to accommodate some minor changes of the original standard. (A new revision is expected in 1997. For that revision, there should be no major changes. Two of the changes are expected to be the elimination of the three automotive companies' logos from the cover and the inclusion of the sanctioned interpretations.)

The introduction of QS-9000 was indeed a Herculean effort to harmonize the automotive quality requirements. It has met the expectations. However, as early as 1995, there were problems about interpretation, intent of several elements within the standard, and applicability to the entire supplier base.

The Automotive Task Force, in order to eliminate these problems, began a systematic effort to answer these questions and concerns. As a result, the *International Automotive Sector Group (IASG) Sanctioned QS-9000 Interpretations* (known simply as sanctioned QS-9000 interpretations) are published on a quarterly basis and the TE Supplement was developed.

The *IASG Sanctioned QS-9000 Interpretations* deals with specific questions, issues, and concerns for which the suppliers need clarification and/or guidance to fulfill the requirements of specific clauses of QS-9000. The answers given in this document are the official interpretations, and they have final authority on the individual clauses of the standard. (A copy of the interpretations may be obtained from the ASQ.)

The TE Supplement was published in July 1996 to reflect the uniqueness of the tooling and equipment industry. The TE Supplement includes the ISO 9001 elements as the base structure of the quality system, adds requirements of QS-9000, and adds or modifies specific requirements applicable only to the tooling and equipment industry. Furthermore, it deletes several elements of QS-9000 that do not pertain specifically to the tooling and equipment industry.

2

ISO 9000 and QS-9000 quality system requirements

This chapter gives an overview of both the ISO 9000 and QS-9000 requirements. In addition, it summarizes key requirements of the standards and offers commentary.

WHAT IS ISO 9000?

ISO 9000 is the generic term for a series of standards sponsored by the International Organization for Standardization (ISO is not the acronym for the organization; rather, it is a Greek word, *iso,* meaning "equal"). These standards specify a generic quality system for all organizations—manufacturing and nonmanufacturing alike. Its intent is to establish, document, and maintain a system for ensuring the output quality of a process. To ensure that this quality system exists, certification is carried out by accredited organizations called *registrars*. These registrars review the facility's quality manual to ensure that it meets the standard and audit the organization's processes to ensure that the system documented in the quality manual is in place and is effective. Once certification is obtained, the certifying body conducts audits of the facility once a year—twice a year with some registrars—and oversees changes and evolutions of the facility's quality system to ensure that it continues to meet the requirements of the standard.

Firms apply for certification under the part of ISO 9000 that most closely fits their form of business. Typically, they fall in one of the following categories:

ISO 9001. This is the most comprehensive part of ISO 9000 and applies to facilities that design, develop, produce, install, and service products or services to customers, who specify how the product or service is to perform. ISO 9001 consists of 20 sections, and it is the part discussed here.

ISO 9002. This standard applies to facilities that provide goods or services consistent with designs or specifications furnished by the customer. It consists of 19 sections of the 20 pertaining to ISO 9001.

ISO 9003. This standard applies to final inspection and test procedures only. It includes 16 of the 20 sections outlined in ISO 9001.

A comparison of the three standards is shown in Table 2-1.

The emphasis of the international standard is on the management of quality systems. While all the clauses of the standard are relevant and important to the company's effective operation, there are certain key features that highlight the strength of the document in terms of correcting the causes of problems and getting all inputs correct with an efficiently functioning process to ensure that outputs are meeting expectations. In essence, the standards are aimed at preventing nonconformance at all stages of a product's "life cycle."

Table 2-1. A Comparison of the ISO 9000 Standards

ISO 9001	ISO 9002	ISO 9003
4.1 Management responsibility	X	X
4.2 Quality system	X	X
4.3 Contract review	X	X
4.4 Design control		
4.5 Document and data control	X	X
4.6 Purchasing	X	
4.7 Control of customer-supplied product	X	X
4.8 Product identification and traceability	X	X
4.9 Process control	X	
4.10 Inspection and testing	X	X
4.11 Control of inspection, measuring, and test equipment	X	X
4.12 Inspection and test status	X	X
4.13 Control of nonconforming product	X	X
4.14 Corrective and preventive action	X	X
4.15 Handling, storage, packaging, preservation, and delivery	X	X
4.16 Control of quality records	X	X
4.17 Internal quality audits	X	X
4.18 Training	X	X
4.19 Servicing	X	
4.20 Statistical techniques	X	X

The standards can be divided into three categories. They are

1. Management responsibilities

   4.1 Management responsibility
   4.2 Quality system

2. Companywide activities

   4.5 Document and data control
   4.8 Product identification and traceability
   4.12 Inspection and test status
   4.13 Control of nonconforming product
   4.14 Corrective and preventive action
   4.16 Control of quality records
   4.17 Internal quality audits
   4.18 Training

3. Specific requirements

   4.3 Contract review
   4.4 Design control
   4.6 Purchasing
   4.7 Control of customer-supplied product
   4.9 Process control
   4.10 Inspection and testing
   4.11 Control of inspection, measuring, and test equipment
   4.15 Handling, storage, packaging, preservation, and delivery
   4.19 Servicing
   4.20 Statistical techniques

SUMMARY REQUIREMENTS AND COMMENTARY OF ISO 9001

In this section, we summarize ISO 9001. We also identify some of the critical issues within each element. For more details, the reader is encouraged to see Stamatis (1995b) and Lamprecht (1992, 1993).

4.1: Management responsibility

Clause 4.1 of the standard emphasizes the responsibility that management has to provide a quality system throughout the organization. In addition, it emphasizes the requirements of a management representative (MR). These requirements include, but are not limited to, facilitating the implementation of the quality system, as well as communicating the system review to management.

Key issues of the review are

the results of internal audits

management effectiveness

nonconformances and irregularities

resolution of customer (internal and external) complaints

solution to quality problems

impact of the current quality system on actual results

4.2: Quality system

Clause 4.2 of the standard is an extension of the management's philosophy and their concern for quality. Specifically, it focuses on a documented system with the main aim to ensure that the organization's products and/or services meet customer requirements.

Part of this quality system is the existence of a quality manual, procedures, and instructions, which include all the requirements of the ISO 9000 standard and the organization's quality policy. The quality manual must support the specific requirements found in the procedures and instructions and must also outline the structure of the documentation used. Minimum requirements of a quality system are

- The preparation of quality plans and documented procedures that concur with the requirements of the ISO 9000 standard and the organization's quality policy
- The identification and acquisition of any controls, processes, equipment, fixtures, resources, and skills that may be needed to achieve the expected quality
- Documented means for identifying customer requirements and for translating those customer requirements effectively into the products and/or services of the entire organization
- The updating process of the quality system
- The modification process and inclusion of new items in the quality system
- A system for evaluating capability

- A defined system of measurement, testing, and control equipment for assessing quality
- A system for defining acceptability
- The preparation of quality records

4.3: Contract review

Clause 4.3 emphasizes the notion that an organization must establish documented procedures for creating, coordinating, and reviewing customer contracts and amendments. Verification of the documented process will focus on the following:

- The customer's requirements.
- Any discrepancies in the contract and how these discrepancies are resolved. Special attention is given to resolution of problems in advance.
- All contract or order conditions are well within the supplier's capabilities.

4.4: Design control

The emphasis of clause 4.4 is on the organization's documentation to control and verify the design of a product in such a way that all of the customer's specified requirements are met. A documented design system should include the following:

- A documented organizational structure that clearly specifies the responsibilities of each design and development activity
- Clearly defined interfaces among the design/development function and its constituencies
- Means to assure that the design/development function has all of the necessary resources and trained personnel
- A system for gathering design input, documenting it, resolving ambiguities, and translating input into the design process
- A system for ensuring that product and service designs meet input requirements, make reference criteria, meet all relevant laws and regulations, and anticipate safety standards
- A system to verify that design output meets all specified design requirements, references acceptance criteria, and identifies design characteristics that are crucial to the safe and proper functioning of the product.

- A procedure for verifying the appropriateness of designs for products and services by means of conducting design reviews, all appropriate and applicable testing, comparative studies, and so on.
- A system for validating designs to ensure that the final product conforms to the defined user needs and customer requirements.
- Procedures for reviewing and adjusting the design and development system as necessitated by circumstances.

4.5: Document and data control

Clause 4.5 reiterates the requirement that the organization administer a documented system for the creation, publication, distribution, use, and revision of all documents and data related to the quality system and the requirements of ISO 9000. A documented control system should ensure that

- Up-to-date editions are readily accessible to anyone who needs them.
- Superseded documents are promptly removed from circulation. (A system for obsolete documents should be in place. These documents can be used to show improvement and certainly to preserve knowledge in the organization.)
- A master list of quality-related documentation and data is maintained, with dates, editions, and issues that are up to date.
- Proper authorization for changes and review procedures exist and are followed.

4.6: Purchasing

Clause 4.6 establishes the need for the organization to establish and maintain documented procedures to ensure that products purchased meet specified requirements. The procedures should provide at least the following:

- All subcontractors must be chosen on the basis of documented capability, past experience, and/or demonstrated ability to meet specifications.
- The extent of control exercised by the supplier over subcontractors must be defined and depends on the type of product being purchased, its impact on the quality of the final product, and past performance of the subcontractors.
- Records of acceptable subcontractors must be maintained.

- All purchasing documents should include data that thoroughly describe the products and/or services concerned, including

  name, type, class, grade, or any other identification

  specifications, drawings, process requirements, inspection instructions, and/or any other relevant data

  the title, number, and issue of the quality standard system to be applied

4.7: Control of customer-supplied product

Suppliers who use customer-supplied products or services in their end products or services must establish and maintain documented procedures for the control and verification, storage, and maintenance of that product. Typical items for consideration are that procedures must exist for ensuring the suitability of the products and services for intended purposes and all products and services (when appropriate and applicable) must be kept safe and secure.

4.8: Product identification and traceability

Identification and traceability are especially vital issues for those organizations that furnish products and services that may be subject to recall if found to be nonconforming, hazardous, or in conflict with laws, regulations, and statutes. It is the intent of clause 4.8 to emphasize the need for establishing and maintaining documented procedures for both identification and traceability throughout all stages of production.

4.9: Process control

According to clause 4.9, the supplier must carry out all production, installation, and servicing processes under a system that specifies thorough planning and control.

A typical process control system must include

- Proper maintenance of equipment to ensure the continued production of quality products
- The approval processes and equipment, as appropriate and applicable

- Documented and updated work instructions specifying the steps required for each task, stipulated in the clearest, most practical manner
- Documented procedures defining the manner of production, installation, and servicing, when the absence of such procedures could adversely affect quality
- Provision of appropriate equipment, facilities, and supplies
- Adherence to standards, laws, codes, quality plans, and/or documented procedures
- Monitoring and control of identified crucial product characteristics at appropriate points in the process

4.10: Inspection and testing

Clause 4.10 confirms the need for establishing and maintaining documented procedures for inspection and testing at each appropriate phase to ensure that the customer's specified product requirements are met. Minimum requirements for inspection and testing are:

- The organization must ensure that incoming product is not used or processed until it has been inspected or otherwise verified as being in conformance with the specified requirements.
- Before the inspection, the organization must establish the amount of control that was exercised at the subcontractor's premises. Evidence of such control must be provided.
- The organization must ensure that all products or services exempted from the receiving inspection procedure, for reasons of urgency, are clearly identified and made traceable so they can be retrieved at any point in the process if there is a need.
- In-process inspection and testing must be clearly documented by the quality plan or documented procedures. The product must be held until the required inspection and tests have been completed or the necessary reports have been received and verified.
- The final inspection process must be carried out in accordance with the quality plan or documented procedures. The process must verify conformance to the specified requirements. No product can be dispatched until all of the activities specified in the quality plan or documented procedures have been satisfactorily completed, documented, and authorized.

- Records must be established and maintained to provide evidence that the product or services have been inspected and/or tested. These records should state clearly whether the product passed or failed the inspection and associated tests. The records must also identify the authority figure who authorized the release of the product.

4.11: Control of inspection, measuring, and test equipment

In order to demonstrate that a product or service is in conformance with its specified requirements, the organization must establish and maintain documented procedures to control, calibrate, and maintain inspection, measuring, and testing equipment. Clause 4.11 of the standards defines that control process. Control procedures must include the following elements:

- The organization must select appropriate and applicable inspection, measuring, and test equipment that meets identified and documented requirements for accuracy and precision.
- Proper identification and calibration with traceability to international or national standards. When no such standard exists, the basis for calibration must be clearly documented.
- The calibration method must be defined and must include details such as

    equipment type
    unique identification
    location
    frequency of check
    check method
    acceptance criteria
    actions taken

- Inspection, measuring, and test equipment must be identified by suitable means, and the records of the calibration status must be retained.
- When calibration is found to be off target, then the organization must assess and document the validity of previous inspection and test results.
- The organization must ensure suitable environmental conditions for all calibrations, inspections, measurements, and tests that are conducted.

- The organization must safeguard the handling, preservation, and storage of inspection, measuring, and test equipment to ensure that it is accurate and fit for future use. In addition, appropriate action has to be taken into account in case adjustments that would invalidate the calibration setting are made.
- In the case of test software, the organization must check the equipment to prove that it is capable of verifying the conformance or non-conformance of a product.
- If technical data are specified requirements for inspection, measuring, and test equipment, the organization must make the data available to the customer so the customer can verify that the equipment is functioning properly.

4.12: Inspection and test status

Clause 4.12 emphasizes that the organization must establish and maintain documented procedures to ensure that the test status of products and services is continuously identified throughout the production process. The system of identification may be visual or it may be organized by physical location or by some other clearly documented means.

4.13: Control of nonconforming product

Clause 4.13 points out the need for the organization to establish and maintain documented procedures for the control of nonconforming products and services. The focus of the procedures is to ensure that such nonconforming products do not inadvertently reach customers. Key points of this section are:

- The control system for nonconforming products must provide for the identification, documentation, evaluation, segregation, and disposition of the nonconforming product. Appropriate notification of the nonconformance to the applicable individuals is necessary.
- Nonconforming products must be reviewed in accordance with pre-defined documented procedures, which may include

  rework to meet the specified requirements
  acceptance with or without repair by concession
  regard for alternative applications
  rejection or scrapping of the nonconforming products

Note: When required by contract, the proposed disposition of the nonconformance must be reported to the customer.

- Records of detection and disposition of nonconformances must be maintained as required.
- Repaired and/or reworked product must be reinspected in accordance with the documented procedures or quality plan.

4.14: Corrective and preventive action

The essence of clause 4.14 is to make sure that the organization has established and maintained documented procedures for implementing corrective and preventive action. The key issues of corrective action are:

- The effective handling of customer complaints and reports of product nonconformances.
- An investigation of the cause of nonconformance relating to the product or service, process, and quality system. The result of the investigation must be recorded.
- The organization must determine what corrective action needs to be taken to eliminate the cause of nonconformance.
- A system of control must be in place to ensure that corrective action is taken and that it is effective.

The key issues of preventive issues are:

- To detect, analyze, and eliminate root causes of nonconformances.
- The organization must determine what steps are needed to handle any problems requiring preventive action.
- The organization must initiate preventive action and apply controls to ensure that it is effective.
- The organization must ensure that relevant information on action taken is submitted for management review.

4.15: Handling, storage, packaging, preservation, and delivery

The organization must establish and maintain documented procedures to ensure the protection of products and services at all phases, from inception through installation. Typical issues of this section include:

- The organization must provide appropriate methods of handling.
- The organization must use secured storage areas to protect products from damage, deterioration, theft, and misuse prior to delivery.
- The organization must control packaging and marking processes to ensure conformance to requirements and prevent unauthorized use.
- The organization must apply appropriate methods for the preservation and segregation of product when it is under the organization's control.
- The organization must ensure that products are protected during delivery.

4.16: Control of quality records

The organization must establish and maintain documented procedures for identifying, collecting, indexing, assessing, filling, storing, maintaining, and disposing of quality records. Quality records must be maintained to demonstrate that a product is in conformance with specified requirements and that the operation of the quality system is effective.

4.17: Internal quality audits

The organization must establish and maintain documented procedures for planning and conducting regular internal audits of the quality system. The purpose of the audits is to determine the degree to which quality activities are being conducted and the related effectiveness of quality activities, areas of nonconformances, and action items.

4.18: Training

The organization must establish and maintain documented procedures for the implementation of a training program to ensure that all personnel can carry out their duties consistent with the objectives of the quality system. For effective training, the following must be considered:

- Quality personnel who are assigned specific tasks. Qualification may be established on the basis of education, training, and/or experience and/or a combination of all three as required.
- Identify skill shortages by means of examination or other techniques.

- Secure the appropriate training resources.
- Verify training effectiveness by means of examination or other techniques.
- Conduct post-training monitoring as appropriate.

4.19: Servicing

In cases for which servicing is a specified requirement, the organization must establish and maintain documented procedures for performing, verifying, and reporting that the service is appropriate to the needs of the customers and the marketplace.

4.20: Statistical techniques

The organization must identify the need for statistical techniques that are required for establishing, controlling, and verifying process capabilities and product characteristics. Even though clause 4.20 is not prescriptive in the definition of the statistical techniques, the organization must establish and maintain documented procedures to implement and control the application of the statistical techniques deemed to be appropriate for the effective operation of the quality system.

THE QS-9000 REQUIREMENTS

The QS-9000 requirements were developed by Chrysler, Ford, and General Motors (1995d). (As of late, the standard has also been adopted by Toyota and Mitsubishi of Australia.) The responsibility for the standard is with the Automotive Task Force; however, its distribution is handled through the Automotive Industry Action Group (AIAG). The intent of this automotive requirement is to harmonize the individual quality standards from each of the companies. For example, QS-9000 replaces Chrysler's *Supplier Quality Assurance Manual*, Ford's *Q-101 Quality System Standard*, and General Motors's *North America Operations Targets for Excellence*.

The core of the standard is ISO 9001, with additional key requirements that fit the automotive industry in three ways:

1. Section I, which is essentially ISO 9001. There are additional requirements in some of the sections, which are shown in italics.

2. Section II is sector specific. These are additional requirements and cover general automotive requirements such as part approvals, continuous improvement, and manufacturing capabilities.
3. Section III is company specific. This section adds specific requirements to the standard based on the specific customer and what the customer has defined as requirements. Typical requirements are the definition of critical characteristics, their symbol, and several others.

In this chapter, we summarize only the additional requirements of QS-9000. The reader should recognize that the summary of the above is valid for QS-9000 as well. Here, we are addressing the added requirements on a per section basis. For more details, see Stamatis (1996b) and Kanholm (1995a, 1995c).

SUMMARY OF THE QS-9000 REQUIREMENTS

Section I: ISO 9001

4.1: Management responsibility

- Organizational Interfaces: Advanced Product Quality Planning (APQP) and control plan manual applies; employ a multidiscipline approach for decision making.
- Business planning: a formal, comprehensive, and documented plan; include operations, finance, new business, quality, and organizational preparedness; both short term and long term. Collect and analyze data to update as necessary.
- Analysis and use of data: use company data to assess company performance, customer satisfaction, goals and objectives.
- Customer satisfaction: documented procedure; benchmarking techniques; management review must happen prior to registration.

4.2: Quality system

- Quality planning: Refer to the *Advanced Product Quality Planning (APQP) and Control Plan* reference manual and *Production Part Approval Process* manual (PPAP): Use of cross-functional teams; feasibility reviews prior to manufacturing; ongoing use of Failure Mode and Effect Analysis (FMEA); control plans shall include three phases: prototypes, prelaunch, production.

4.3: Contract review

No additional requirements.

4.4: Design control

- Design and development planning: specific skill qualifications as appropriate.
- Design input: computer-aided resources are needed as appropriate; CAD/CAE (Computer Aided Design/Computer Aided Evaluation) system capable of two-way interface with customer systems.
- Design output: the process shall include various design optimization and analysis tools.
- Design verification: must include a prototype program; performance tests shall include product life, reliability, and durability analyses as appropriate.
- Design changes: customer approval required per the PPAP manual and QS-9000.
- For proprietary designs, consider form, fit, function, performance, and/or durability.

4.5: Document and data control

- Reference documents: all customer-referenced documents shall be available; procedure must include implementation of customer design changes.

4.6: Purchasing

- Approved materials for ongoing production: purchase only from approved subcontractors; all material must comply with the current government and safety constraints.
- Subcontractor development: suppliers shall perform subcontractor quality system development; QS-9000 assessment can be accomplished by the customer, an approved original equipment manufacturer (OEM) second party, or an accredited third-party registrar;

suppliers shall require 100% on-time delivery and monitor this performance.
- Restricted substances: comply with regulatory and safety constraints.

4.7: Control of customer-supplied product

No additional requirements.

4.8: Product identification and traceability

No additional requirements.

4.9: Process control

- Government safety and environmental regulations: process and procedures shall include compliance with safety and environmental regulations.
- Designation of special characteristics: designate and document the control of special characteristics.
- Preventive maintenance: identify key processes requiring preventive maintenance; document and plan the activities based on process/equipment performance.
- Process monitoring and operator instructions: prepare operator instructions; include all pertinent operating instructions.
- Preliminary process capability requirements: preliminary process capability studies are required; set limits of acceptability (PPAP manual).
- Ongoing process performance requirements: apply Statistical Process Control (SPC) for performance evaluation; continually improve, especially special characteristics; modified preliminary or ongoing capability requirements per control plan.
- Verification of job set-ups per documented instructions: process changes require customer approval; appearance items must be manufactured with proper controls; workmanship standards.

4.10: Inspection and testing

- Acceptance criteria: zero defects; visual standard acceptance approved by customer.
- Accredited laboratories: used on direction of customer.

- Incoming product quality: incoming quality can be assessed by various techniques; review of data, inspection, audits of subcontractors, third-party testing, and certifications.
- Layout inspection and functional testing: apply customer standards; frequency established by customer.

4.11: Control of inspection, measuring, and test equipment

- Inspection, measuring, and test equipment records: records must include revisions following engineering changes, gage conditions and actual readings as received for calibration/verification, and notification of customer if suspect material has been shipped.
- Measuring system analysis: variation analysis via appropriate statistical techniques; apply the *Measurement Systems Analysis* reference manual.

4.12: Inspection and test status

- Product location: location alone may not suffice to identify nonconforming product unless inherently obvious.
- Supplemental verification: as required by customer.

4.13: Control of nonconforming product

- Suspect product: this element applies to suspect product, as well as to nonconforming product.
- Control of reworked product: instructions shall be provided; reduction programs shall be implemented; no visible rework on parts and materials for service applications.
- Engineering approval product authorization: approval is required for rework via unapproved processes.

4.14: Corrective and preventive action

- Problem-solving methods: use disciplined problem-solving techniques; for external nonconformance, respond as required by customer.
- Returned product test/analysis: all returns shall be analyzed; process modifications shall be implemented as needed.

4.15: Handling, storage, packaging, preservation, and delivery

- Inventory: inventory management shall be utilized to optimize inventory turns, ensure stock rotation, and minimize inventory levels.

- Customer packaging standards: apply customer standards and/or guidelines; apply required labels.
- Supplier delivery performance monitoring: strive for 100% on-time delivery; deliver per customer requirements; advance shipment notifications shall be utilized.

4.16: Control of quality records

- Record retention (unless otherwise specified or regulated): maintain during active production and service life, plus one year; part approvals, tooling records, purchase orders; quality performance for one year after the year of creation; audits and management reviews for three years; superseded parts
- Copies of pertinent documents shall be upgraded to new parts records.

4.17: Internal quality audits

- Inclusion of working environment: suitable working environment shall be considered; entire system must be audited prior to QS-9000 registration audit.

4.18: Training

- Training as a strategic issue: training affects all personnel; effectiveness shall be periodically evaluated.

4.19: Servicing

- Feedback of information from service: a procedure to address service concerns is required; provide feedback to manufacturing, engineering, and design as needed.

4.20: Statistical techniques

- Selection of statistical tools: shall be determined during advanced quality planning.
- Knowledge of basic statistical concepts: should be known throughout the organization as appropriate; apply the *Fundamental Statistical Process Control* reference manual.

Section II: Sector-specific requirement

• Production Part Approval Process. Full compliance is required to the PPAP manual. Any questions are to be referred to approving authority. Production parts are approved by part number and revision numbers by site manufactured, including subcontractors and materials used. Engineering change validation is the supplier's responsibility.

• Continuous Improvement. It is expected that suppliers will fully deploy a philosophy of continuous improvement. Improvement action plans are to be written for critical processes. Improvements to characteristics measured by attribute should focus on perfection of process methods. Productivity and efficiency programs will be implemented to reduce downtime, reduce setup time, and trace causes of downtime. Examples of continuous improvement processes and techniques are listed in this section.

• Manufacturing Capabilities. Methods must be documented for analysis of facilities and plants within the APQP manual. Mistake-proofing the process is to be addressed in the planning process. Appropriate technical resources are to be used in the design and fabrication of process tooling. The supplier is to employ a tooling management system including maintenance and repair, storage, setup, and tool changes system for perishable tools.

Section III: Company-specific requirements

• Section III of QS-9000 covers company-specific requirements. They include drawing designations for special characteristics, inspection or layout frequencies, labeling requirements, materials analysis, and so on, and details that need to be reviewed as part of the implementation plan, but are dependent on the supplier customer base.

It is of paramount importance that the reader keep up with the official interpretations of the standards. The official interpretations are published periodically by the Automotive Task Force under the name *Sanctioned QS-9000 Interpretations*. The intent of these interpretations is to clarify each of the elements of QS-9000 as they apply to the suppliers. These interpretations may be obtained from the American Society for Quality (ASQ).

3

The TE supplement quality system requirements

This chapter introduces the TE Supplement and discusses the issues concerning the tooling and equipment industry.

WHAT IS THE TE SUPPLEMENT?

The TE Supplement is an extension of QS-9000. It was designed by Chrysler, Ford, General Motors, and Riviera Die and Tool with the intent of clarifying the quality system (QS) requirements to the tooling and equipment supplier base. All tooling, equipment, and machinery suppliers are to follow the QS-9000 requirements and the TE Supplement when applicable. Typically, suppliers dealing in the following processes must adhere to the TE Supplement:

stamping	painting	heat treating	balancing	tooling
molding	assembly	welding	casting	measuring
deburring	gaging	forming	forging	packaging
robotics	plating	washing	packing	machining

To determine whether or not you are a tooling and equipment supplier, a good rule of thumb is the following: If you manufacture (produce) a product for the automotive industry, but it does not travel down the road with the vehicle, more than likely you are a tooling and equipment supplier; for example, machine suppliers that produce a conveyor line or manufacture paint equipment. The supplement was created to harmonize Chrysler's *Tooling and Equipment Supplier Quality Assurance* requirements, Ford's *Facilities and Tools Quality Systems Standard*, and QS-9000.

As with QS-9000, this new supplement contains ISO 9001 in its entirety, along with additional customer-specific requirements and those requirements specifically geared toward the tooling and equipment industry. This means that any nonautomotive customers' requirements for ISO 9001 will also be met with third-party registration to the TE Supplement requirements; companies that obtain third-party registration are officially ISO 9001 certified. Of course, any company may go beyond the scope and the requirements of the supplement.

Suppliers are required to devise, record, and put in place an effective quality management system according to their customers' and the TE Supplement requirements. It is imperative that all the elements of the TE Supplement be included in their quality management system and must be defined in the supplier's quality manual. In turn, the quality manual must be used as an input document when developing the quality system.

TE SUPPLEMENT REQUIREMENTS

In order to meet the requirements of the TE Supplement, all of the standard's elements must be included in your quality system and corresponding documentation. Just as in the ISO 9000 and QS-9000 structure, the supplier that does their own design must follow all the requirements of ISO 9001. On the other hand, the supplier with products that are designed and serviced by a subcontractor can use the ISO 9002 quality system model. This same rationale holds true for tooling and equipment companies. That is, they should follow the requirements of ISO 9001 if their design is done within the facility, or they should follow the requirements of the ISO 9002 quality system model if there is no design performed within the facility. The ISO 9002 model excludes the design control element found in ISO 9001.

In the following, we address the additional and unique requirements of the TE Supplement. The ISO 9001:1994, Section 4, and QS-9000 requirements are addressed in Chapter 2.

Section I: ISO 9000

4.1: Management responsibility

4.1.1: Quality policy
   • Quality policy. Quality objectives shall contain reliability, maintainability, and durability.

4.2: Quality system

4.2.3: Quality planning

• Quality planning. When a new requirement replaces an entire section of QS-9000, the ISO 9001 portion remains intact. The use of advanced quality planning is mandatory. Understanding of quality/reliability requirements of the customer is mandatory. Controls shall be designed and executed throughout the concept, design and development, prototype, machinery build, and test phases to make sure that the requirements for quality and reliability are being followed and documented.

• Use of cross-functional teams. The *Reliability and Maintainability Guidelines* (R&M Guidelines) shall be used as the reference in the implementation of R&M methodologies.

• Feasibility reviews. Feasibility reviews must be documented.

• Failure Mode and Effect Analysis (FMEA). The following requirements replace the QS-9000 requirements. FMEAs shall be documented, but also may be used in generic processes or family of products when appropriate. Recommended sources for conducting an FMEA are the FMEA Manual and the R&M Guidelines. Internal processes should be addressed using this discipline. FMEAs should be living documents, reviewed and updated.

• Control plan. The section on control plans replaces the QS-9000 control plan section. Control plans must be prepared by the company to address engineering requirements, special characteristics, and process controls for the manufacture of tooling and equipment. The *Advanced Product Quality Planning and Control Plan Manual* is suggested as a model for control plans. Control plans may be based on existing plans for mature products. Control plans shall be living documents and shall be reviewed and updated. Customer approval may be required.

4.3: Contract review

No additions.

4.4: Design control

4.4.2: Design and development planning

• Required skills. In addition to the skills identified in QS-9000, this section identifies the following as required skills: mean time to repair

(MTTR); mean time between failures (MTBF); fault tree analysis (FTA); and life cost cycle (LCC). Guidance for these skills may be found in the R&M Guidelines.

4.4.5: Design output

• Design output—supplemental. Design output must include analysis of test data and projections of reliability, maintainability, durability, and life cost cycle.

4.4.7: Design verification

• Design verification—supplemental. The following replace the QS-9000 requirements. Performance tests must include appropriateness, maintainability, durability, reliability, and product life cycle. A comprehensive prototype program must exist, and the company must utilize predictive R&M techniques. Accelerated life testing must be conducted on crucial components of machinery or equipment.

4.4.9: Design changes

• Design changes. The company must maintain a log of all design changes through each phase of the machinery build.

4.5: Document and data control

No additions.

4.6: Purchasing

4.6.2: Evaluation of subcontractors

• Subcontractor development. All the requirements of QS-9000 apply, as well as those of the TE Supplement, Sections I and II.

4.7: Control of customer-supplied product

No additional requirements.

4.8: Product identification and traceability

• Product identification and traceability. The company must establish and maintain a tracking system for components and subassemblies. The

system shall include identification of the components, job nu
reference engineering drawings, and bill of materials.

4.9: Process control

4.9.1: Process monitoring and operator instructions

• Process monitoring and operator instructions. This section replaces the QS-9000 requirement; however, the ISO 9001 requirements remain in their entirety. Appropriate and adequate process monitoring and operator instructions must be available to document the tooling and equipment process. Typical items are process sheets, inspection and laboratory test instructions, shop travelers, test procedures, standard operation sheets, and a control plan.

4.9.2: Initial process study requirements

This section replaces the QS-9000 requirements; however, the ISO 9001 requirements remain in their entirety. Initial process studies are required for each supplier- or customer-designated special characteristics for internal processes. The data must meet customer requirements. Appropriate reference is given in Section II and III of the TE Supplement.

4.9.5: Verification of job setups

Does not apply to tooling and equipment companies.

4.9.6: Process changes

Does not apply to tooling and equipment companies.

4.9.7: Appearance

Does not apply to tooling and equipment companies.

4.10: Inspection and testing

4.10.4: Final inspection and testing

• Layout inspection and functional testing. This section replaces the QS-9000 requirements; however, the ISO 9001 requirements remain intact.

Functional verification must be performed for all tooling and equipment. Specific requirements are listed on page 20 of the TE Supplement.

4.11: Control of inspection, measuring, and test equipment

No additions.

4.12: Inspection and test status

No additions.

4.13: Control of nonconforming product

4.13.3: Control of reworked product

Does not apply to tooling and equipment companies.

4.13.4: Engineering approved product authorization

Does not apply to tooling and equipment companies.

4.14: Corrective and preventive action

No additions.

4.15: Handling, storage, packaging, preservation, and delivery

4.15.6: Delivery

• Supplier delivery performance. This section replaces the QS-9000 supplier delivery performance monitoring section. The company must establish a 100% goal for delivery through timing plans and critical path scheduling. The company should develop resource plans that identify capacity, labor, and facilities. The company must have in place scheduling systems that accurately control start and end timing for the manufacture of all major components and assemblies, for equipment tests, and for runoff, installation, and tryout. Subcontractors should also have scheduling systems. Meetings should be held to determine the status of actions to maintain timing and to determine the effect of engineering changes.

• Production scheduling. Does not apply to tooling and equipment companies.

• Shipment notification system. Does not apply to tooling and equipment companies.

4.16: Control of quality records

No additions.

4.17: Internal quality audits

No additions.

4.18: Training

• Training. The company must have a formal training program. Must include R&M.

4.19: Servicing

• Servicing. A procedure for communication of information on machine uptime, reliability, maintenance history, and service must be established and communicated.

4.20: Statistical techniques

4.20.2: Procedures

• Knowledge of statistical concepts. In addition to the requirements of QS-9000, the following are included: MTBF; MTTR; short-run SPC; control charts; and p, np, c, and u charts or any other appropriate statistical technique. The company should ensure that all personnel understand basic concepts such as capability, variation, stability, and overadjustment.

Section II: Sector-specific requirements

The Production Part Approval Process does not apply to tooling and equipment companies. However, this section has been replaced with the following:

1.0: Purpose

The purpose is to develop a system that ensures robots, machinery, tools, and equipment purchased by a customer will be of acceptable quality in both function and reliability when received.

1.1: Machinery qualification runoff requirements

Root cause analysis for all failures must be documented. Specific requirements for supplier location and customer plant. In summary format, these requirements are shown in Table 3-1.

Table 3-1. Summary Requirements for Machinery Qualification

Location	Supplier presentation	Time set for production rates	Items selected	Requirements
At supplier	50/20 dry run	50 hours for robots 20 hours all other	None	Production operation without failures and/or breaks
	Preliminary. evaluation	As required	As required	Stability demonstrated and documented
	Supplier stability and capability	As required	As required	Consult Appendix I of the TE Supplement
	Proof of performance through P_{pk}	As required	As required	As per purchase agreement or supplier target
	Reliability verification	Not applicable	Not applicable	As per purchase agreement or supplier target
At customer	20-hour dry run	20 hours	None	Production operation without failures and/or breaks at specified speed
	Short-term performance	As required	As required	As per purchase agreement
	Long-term performance	25 working days	As required	As per purchase agreement

1.2: Procedure

• 50/20 run. (The approach and application of the 50/20 run are summarized in Table 3-1.) The 50-hour test is for robots. The 20-hour test is for all others. The goals of the 50/20-hour dry run may be summarized by the following: reduce start-up time; resolve software and control problems; verify the machinery's reliability; confirm and verify the equipment cycle to the requirements of the customer; and improve the quality of all sections.

2.3: Continuous improvement

• Techniques for continuous improvement. In addition to the QS-9000 requirements, there must be demonstrated knowledge in the following measures: MTBF; MTTR; LCC; and reliability growth.

Section III: Company-specific requirements

Chrysler

- Third-party registration requirements. As of this writing, third-party registration is not required.
- Lot acceptance sampling table. In most cases, this item does not apply. When it does, the requirements of QS-9000 will prevail.
- Product qualification. In most cases, this item does not apply. When it does, the requirements of QS-9000 will prevail.
- Equipment qualification guide. The TE Supplement provides the user with a table of process performance Ppk with specific guidelines for its use.

Ford

- Third-party registration requirements. As of this writing, third-party registration is not required.
- R&M program planning requirements. Includes MTBF; definition of failure; R&M plan; design assurance strategies; design reviews; FMEA; reliability testing and assessment; and R&M continuous improvement activities (a table is provided as a guide for the user).

General Motors

- Third-party registration requirements. As of this writing, third-party registration is not required.
- Training. Designers and engineers must be trained in R&M.
- Design review. Document the R&M through the program planning worksheet. Appropriate documentation is considered MTTR, MTBF, reliability block diagrams, and design review.
- FMEA. FMEA must be used on components and subsystems.
- Maintainability requirements. Additional requirements are accessibility, diagnostics, maintenance procedure, suggested spare parts inventory, and appropriate training.
- R&M validation. Must be conducted at the supplier's site based on General Motors Master Test List. Must be demonstrated at GM site using GM's criteria. Appropriate validation must be used at installation.
- Continuous improvement. Supplier must participate in data collection and problem resolution. The following documents must be followed by the tooling and equipment suppliers: *General Motors Corporation Standard for New and Rebuilt Machinery and Equipment,* GM Specification No. V 1.0-1993 (or latest version) (GM-1761); *General Motors Corporation Laser Alignment Specification for New and Rebuilt Machinery and Equipment,* GM Specification No. A 1.0-1993 (or latest version) (GM-1907); infrared inspection, not completed.

4

Implementation strategy

This chapter discusses the implementation process of the TE Supplement in a given organization. Specifically, a discussion is presented about the prerequisites to implementation, followed by two alternative approaches for implementing a quality system, the role of management, and the benefits of the TE Supplement.

PREREQUISITES TO IMPLEMENTATION

There is no single, agreed upon approach to implementing a quality system. However, there are some issues and concerns that all companies must address when embarking on an implementation program. Typical issues are

- work structures
- configuration of practices
- configuration of processes

To be successful in implementing a quality system, such as the TE Supplement, it is of paramount importance to develop first the team concept through cross-functional and multidisciplinary members, and second, to empower these individuals within their work environment to have both appropriate and applicable authority and responsibility for actions taken concerning their practices and processes. This, of course, leads to ownership and eventually enables the employees to work to their full potential and toward the same goals and vision of the organization.

How is this done? To be sure, all organizations are complex systems, and some are more complex than others. However, for success to occur, it is imperative that there be an agreement among the communication flow of understanding throughout the organization, consistent performance, and goals that are in congruence with its business environment. Some of the elements for doing precisely this are shown in Figure 4-1; they are:

- Create a vision, mission, and goals
- Articulate the beliefs and values
- Define the management practices
- Design the appropriate and applicable organizational structure
- Define and design appropriate and applicable work practices and processes
- Define and design appropriate and applicable human resources systems
- Identify and incorporate appropriate and applicable technology throughout the organization

Strategic agreement

At the highest level of the company, the business strategy of the organization must be in tandem with its business environment, vision, mission, beliefs, and values. Therefore, selecting a business strategy is important because it guides the goals of the entire organization. However, it must be emphasized that not all goals can be pursued at the same time with the same vigor and priority.

On the other hand, beliefs and values become clearer when organizations make conscious tradeoffs among conflicting goals. Without clear and defined guiding principles, financial goals usually take precedence over other considerations. Therefore, it is imperative for any organization to spend time to articulate its values so that a balance among a variety of objectives may be reached. Examples of conflicting objectives may be quality performance, customer satisfaction, financial results, environmental requirements, and so on.

Goal agreement

Once the organization has reached an agreement between goals and business strategy, it must specify these goals for different units and processes. The idea here is to focus and streamline individual goals to individual processes on one hand and on the other hand to maximize the organizational goals as well.

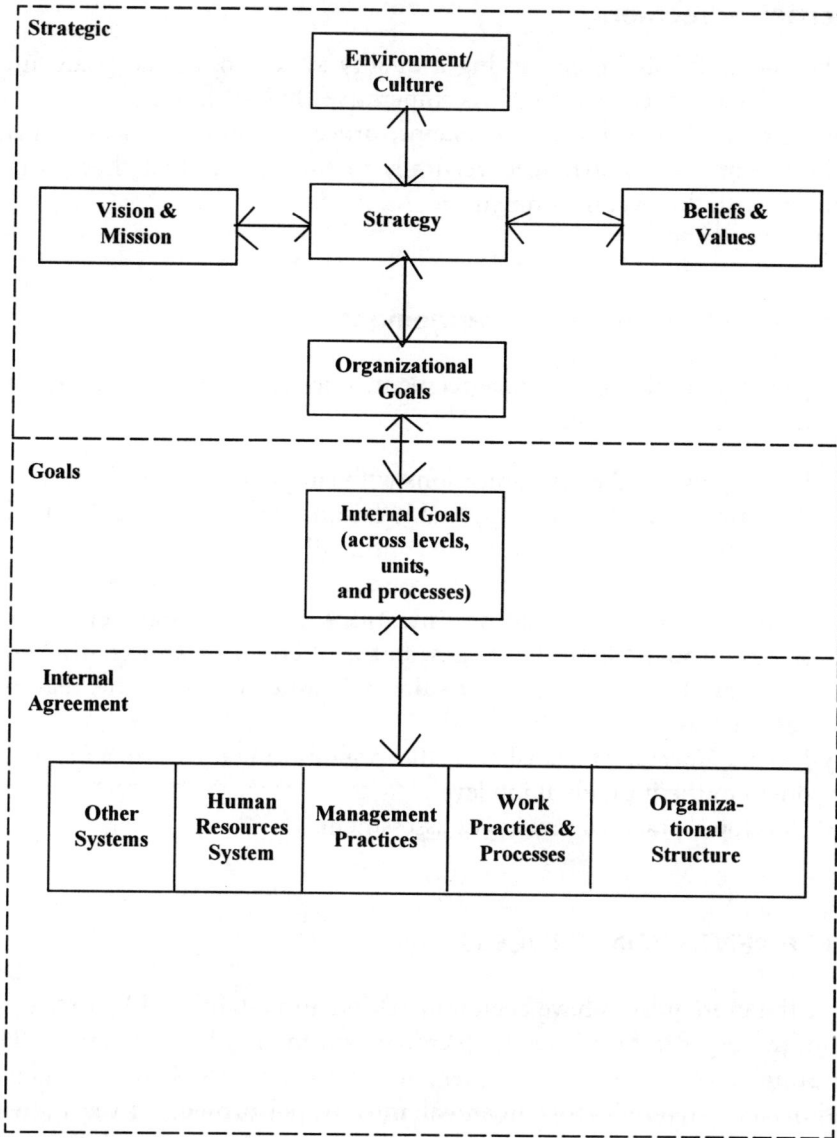

Figure 4-1. Model for agreement.

Internal agreement

To be successful in implementing a quality system once the goals and objectives have been defined, it is imperative that all internal organizational elements—structure, subsystems, processes, and practices—must be in agreement and arranged vertically in such a way that there is no problem with the overall company goals.

Critical factors for this arrangement

To help maintain this system of agreement, as well as arrangement, the following factors are critical:

- Develop external communications with suppliers.
- Allocate appropriate and applicable resources to make sure that both agreement and arrangement are continually monitored.
- Develop flexibility to adapt new conditions, environment, and so on.
- Encourage continual communications with internal personnel. Always notify them of changes, status of changes, and result(s) of changes. Encourage them to do the same, without the fear of retaliation.
- Provide means of translating the organizational vision and goals down to the individual job level.
- Develop a practical process of assessment.

IMPLEMENTATION PROCESS

Once the prerequisites have been understood and established by management, the organization is ready to embark on the implementation of the TE Supplement. The actual implementation process is defined in eight critical steps. All are important and all must be performed. However, their sequence can vary. In fact, quite often in practice they overlap; the early steps are always dynamic—quite often they change due to changes in the process. The critical steps are given below.

Step 1. Build the case for change. To set the context of change, management must be willing to educate everyone in the company about the state of the industry, competitors, customers, and other factors that drive the business. In other words, management must strive for "business literacy" at all levels. They must assess the company's capabilities, culture, and readiness for change through surveys, internal audits, and gap analysis. They must communicate the performance and capabilities needed to compete. The operative issue is "needed to compete."

Once the case for change has been made, then management must move quickly and consistently to implement it effectively in a "top down" fashion. This may be accomplished through meetings, training, conferencing, newsletters, testimonials, speeches, and so on.

Step 2. Define vision, mission, and strategy. Unless the company defines its vision, mission, and strategy, it is going to have problems down the road. One of the best ways to define the vision, mission, and strategy is to create a steering committee.

Step 3. Develop a communication and involvement strategy. This is perhaps one of the most important steps in any organization. The function here is to develop a system, a plan for engaging people at all levels of the company in the implementation process. Part of the system should be to communicate the current state, desired future state, transformation process, expectation results of proposed change, and a plan to deal with rumors.

Step 4. Design and/or redesign the organization. The idea of this step is to find out what is the current status. Conduct a needs analysis to identify the "gaps" of your organization and then proceed either to design a new process or to redesign the current process and develop appropriate and applicable documentation. In this step, it is imperative to involve people who actually are doing the work.

Step 5. Plan for implementation. Key issues of this step are:

- Reaffirm management commitment
- Assess readiness for implementation
- Develop ownership commitment
- Create an implementation structure
- Identify and prioritize key processes
- Develop a transition plan and structure with measurable milestones that support the change
- Clarify the implementation team's boundaries

Step 6. Implement the system. In this step, management must maintain a visible role and ensure continual support for the implementation process. One of the ways that this is accomplished is by making sure that all appropriate resources are available and on time. Deliverables of this step are the quality manual

procedures

instructions

records

Step 7. Monitor and evaluate progress. Perhaps this step is the easiest, primarily because part of the implementation process of a quality system is the notion of building a mechanism for feedback. This, of course, is part of the TE Supplement requirement in clause 4.17, which concerns the internal audit. It is through this element that the monitoring and evaluation can produce information that can help both process operators and management to improve.

Step 8. Renew the commitment. Once the implementation of the quality system has been accomplished, it is imperative that management designate a champion to maintain the integrity of the system. This is usually done with the management representative (MR). After the completion of the implementation, this champion serves as the person in charge of continual analysis of the system, appropriate utilization of the quality system, and, more importantly, working on future and/or next-generation changes and the prevention of backsliding.

AN ALTERNATIVE IMPLEMENTATION MODEL

There is no "magic" and/or "best" way of implementing a quality system for *all* organizations. The best is and must be defined by the individual organization pursuing the implementation of a quality system. In this section, we provide the reader with yet another methodology, this time based on a six-step approach.

Step 1. Create an organizational structure, including the selection of the MR. This step is based directly in the standard (TE/QS-9001—4.1.2.3). According to this section of the standard, the MR is the person within the facility who acts as a liaison or an "interface" between the facility's management and everyone else. The overall structure must be conducive to plan and oversee the overall implementation process.

Step 2. Formulate a steering committee. This step involves management. This top-level implementation team (action council) should include the chief executive officer, top managers, key functional managers, top-level and key union representatives, and the champion of the program. The council's function is to serve as a policy group, that is, set objectives, approve plans, allocate resources, evaluate reports, allocate priorities to needs, and make changes (as needed) during the implementation process.

Step 3. Perform a needs assessment of your current quality system. This step serves as an exploratory investigation of the organization's current position. The idea here is to find out the organization's position in comparison to QS-9000 expectations. The differences, obviously, are the gaps and the areas in which the organization should focus—at least in the short run. This assessment may be conducted by internal qualified personnel or outside consultants. In either case, the main purpose of the assessment is to get a clear picture of the state of the facility's quality program as it compares with the TE Supplement and QS-9000 requirements.

Step 4. Train personnel. The need for training can vary from organization to organization. However, training should be looked on as a strategic advantage for the entire organization. Specific training should be identified and delivered on an "as needed" basis. This need should meet the gaps identified in Step 3, and it should also provide functional training for specific designated areas of operation. In other words, training should be provided throughout the organization to meet immediate, as well as strategic, goals.

The immediate goals should cover generic training in the entire facility for awareness, expectations, and benefits of this implementation program. The strategic goals, on the other hand, should cover specific training in areas such as advanced quality planning, Failure Mode and Effect Analysis (FMEA), Reliability and Maintainability (R&M), and so on. Both the immediate and the strategic goals should encourage employees to support and demonstrate commitment and enthusiasm toward the project and its success.

Step 5. Set up an implementation team. The idea here is to identify the "doers" of the implementation process. These individuals (in team formation) will carry out the quality policies devised by the council and are responsible for putting the quality system into action. It is imperative that these teams should be made up of individuals who are knowledgeable about the standards and culture of the organization and have some training in group dynamics, as well as the processes to which they are assigned.

As a rule, these teams are organized by department or function and are typically led by a functional manager or a department head who sits on the steering council. In any case, the overall objective of the team is to include enough representatives of each of the organization's critical functions.

It is these teams that will eventually write the quality manual, procedures, and instructions of the quality system. Therefore, it is very important that these individuals are trained appropriately and know the organization and its practices quite well.

Step 5a. Hire an outside consultant (optional). This step is an optional one; however, if it is chosen by an organization, you should expect the same results as in Step 4. The major difference between the in-house teams and the consultant is that, at the end of the process, organizations who have done it "on their own" will have ownership and understanding of what the quality system is all about; as a consequence, success is imminent. If a consultant is to be retained, we strongly recommend that the consultant's function should be that of a coach, director, troubleshooter, supporter, and, perhaps, planner. No organization should hire an outside consultant to do the "whole thing." It will not work in the long run.

Step 6. Create an implementation plan that can be documented. In Step 5, we mentioned the importance of teams in the implementation process of any quality system. This is important because the process of creating and using documentation is the core to the effectiveness of any quality system implementation. After all, you cannot have one without the other.

Documenting the quality system is also educational as it forces the facility, at all levels, to think through exactly what is being done and how it is being done. Documentation of the quality system is mandatory; there is nothing anybody can do about it except to have documentation that is representative of what the organization "is doing" and not what the organization "would like to do."

To prove that the implementation plan can be documented, a system must be developed that proves the accuracy, effectiveness, and integrity of the system. That system is usually in three or four tiers. Level 1 is the quality manual, a document describing the existing quality system. Level 2 is the procedures level, which defines the way to perform an activity and how the quality system works. Level 3 is the instruction level, which gives specific directions or the detailed requirements for performing an activity listed under the procedure. Sometimes, as part of the instructions, records may be incorporated in this level as well. Level 4 is the records level and can contain drawings, forms, files, technical standards, and so on that will support the quality activities of the facility. Chapter 6 elaborates on the documentation issues of a quality system.

THE ROLE OF MANAGEMENT

With every program, some companies are further along than others in both effectiveness and implementation. The TE Supplement effort is no different. To make sure that a successful program within the bounds of the TE Supplement takes place, we recommend that senior management must be the leaders in communicating the need for change and its significance to the company. They must set a clear direction and reinforce their messages by allocating the appropriate and applicable resources, as well as realistic times of implementation. They must also understand that things are likely to go astray if they do not reinforce the idea of a quality system to the middle managers. In essence, they must be committed.

To say that management is committed is very easy; however, to demonstrate commitment to all employees is much harder. If you, as a manager, can prove that you do the following, you are well on your way to demonstrating commitment.

First, have the basics of quality operating systems in your process: understanding of quality policy; documentation of machines, materials, manpower, processes, and so on; process flow chart; APQP implemented; process FMEAs; control plans; SPC (as applicable) of your processes; training spelled out and documented; process control methods; and feasibility studies and reviews.

Second, show and document your commitment to quality product; your commitment to scrap, rework, and reject tracking; your improvement in performance; your system for tracking other key measurables; your use of ideas from, and recognition of, employee participation; SPC use with documentation of reactions to out-of-control instances; your commitment to auditing your progress; your commitment to improvements in process; your commitment to safety, health, and government regulations; authorization and delegation to appropriate personnel; your organizational skills, both personal and organizationwide; and your system outline so others can find and do things in your absence.

Third, convey to all employees the following: quality practices; safety practices; written definition of responsibilities (what to do, when to do it, how to do it, where to do it, why it needs to be done); and written definition of authorities (what you are authorized to okay, what you need a "boss" to authorize as okay, whom you see if the boss is not present to authorize an item).

Fourth, handle communication as follows. Hold meetings to see how things are going (right or wrong); you will need agendas and minutes, people from your area plus outside people as they are affected, participation of appropriate personnel, and volunteers noted if used. Problem-solving teams must have agendas, minutes, and root cause analysis and must prevent reoccurrence, plan and do the containment, plan and do the corrective action, and be the decision-making unit (have authority). For customer-related issues, you must participate in APQP; get involved in delegating and authorizing subordinates to specific projects (always follow up on the results); be aware of changes and communicate them in the organization; and expect documentation of problems, with all solutions communicated to the areas affected. For supplier-related issues, you must participate in APQP, actively participate in supplier relations either in their development or in their quality endeavors, encourage your subordinates to participate actively in supplier relations and always demand follow-up, be aware of changes and communicate them to all appropriate personnel, expect documentation of problems and their outcomes, expect all problems to be addressed at the level they occur, check the system through internal auditing and gap analysis, and expect proof and documentation of critical activities.

Middle managers play a critical role in the implementation process since they are indeed the ones who control the resources, including the time of the employee. Unless they are on board, the implementation process is going nowhere.

One of the ways to improve the process of implementing the TE Supplement is for management to use the concept of project management. The project manager is the appropriate, and perhaps the only, properly qualified professional to speed up the cumbersome communication paths of the hierarchical structure typical in most organizations. For a very detailed approach of implementation and the utilization of project management, see Stamatis (1996).

THE BENEFITS OF CONFORMANCE TO THE TE SUPPLEMENT

Companies who pursue the TE Supplement will be categorized as companies who are committed to quality excellence. Furthermore, if, at a future date, third-party registration is required, it will offer great exposure to the world markets to the company receiving the certification.

Just as QS-9000 was introduced by the automotive companies to harmonize most of their quality standards, the TE Supplement has the same goal. By harmonizing the standards for the tooling and equipment industry, the benefits of that standardization will result in a lower cost to develop and maintain individual quality standards.

In the 1990s, the "mantras" of quality are continuous improvement and customer satisfaction. Toward this end, the TE Supplement will ensure that the customer is getting goods and services that are needed, wanted, and expected. This straightforward approach of satisfying the customer is one way to build trust between your company and the supplier. In addition to this trust, a company pursuing the TE Supplement standardization will improve internally as the employees achieve and maintain the discipline of conformance, which is what the customer really wants.

Finally, one more benefit that a company may realize is financial. Companies that pursue the requirements of the TE Supplement will save money because they now have to prepare documentation for only one quality system to meet all of their customers' needs, as well as eliminate waste from their processes.

5

Documentation, auditing, and certification

This chapter gives an overview of documentation, auditing, and certification. Specifically, the documentation issues of a quality manual, procedures, and instructions are addressed. For auditing, a general overview is given, as well as the TE approach. Finally, the issues of certification are discussed. The reader is encouraged to see Stamatis (1996a) for a detailed explanation of documentation and auditing as it applies to both ISO 9000 and QS-9000.

DOCUMENTATION

Clause 4.2 of ISO 9001 requires companies to "establish, document, and maintain a documented quality system as a means of ensuring that product conforms to specified requirements." Therefore, when developing a quality system, it is imperative to develop it in such a way that embodies all the elements of the standard, as well as the policies, procedures, and instructions of the company.

A typical documentation system is one that reflects these requirements; for simplicity only, it is divided into the following levels:

Level	Documentation	Document
Level 1	Policy	Quality policy manual
Level 2	Procedures	Procedures manual
Level 3	Work instructions	Instruction manual
Level 4 (optional)	Forms, tags, and labels	Instruction manual

The quality manual

The first documentation level is what is generally called a quality manual, but it essentially deals with policy and describes the way in which the company system is structured to meet the requirements of the quality management standard. When developing the quality manual, it should be remembered that the primary purpose of a quality manual is to provide an adequate description of the company's quality system. At the same time, it serves as a permanent reference for the implementation and maintenance of that system. For precisely this reason, we strongly recommend that an efficient and effective method of updating is provided to allow the system to change over time.

The structure of the quality manual can vary from company to company, although many organizations currently develop the quality manual, or first-tier documentation, in the order of the requirements included in the standard. In this way, the policy manual effectively provides a detailed index to the quality system of a company.

The reason for such a sequence is that it makes the quality manual somewhat easier to audit; however, that is not the main criterion for structuring the document. It should be produced to meet the needs of the user rather than the registrar and/or the auditor. In essence, the quality manual should reflect what the company does rather than provide an indication of what it would like to do. For an example of a quality manual, see Appendix A.

Quality procedures and work instructions

All organizations exist to do "something"—either to produce a product or to deliver a service. However, in order for organizations to achieve that goal, they must follow some process of procedures and/or instructions. Therefore, the second and third tiers of documentation are the procedures and instructions, respectively. No one will deny that procedures and instructions exist "someplace" in the organization. That is not the point of the ISO 9000 documentation. The point is that the documentation must exist for everyone involved with a specific task to know what that task is, how it is done, and what to do if something goes wrong. The procedures and the instructions tell the user in a detailed form how to approach their job in a structured manner. This structure, however, allows the system some flexibility; in fact, flexibility is built in—when a change is required, there are procedures and instructions for incorporating that change in the system. For a variety of examples of both procedures and instructions, see Appendix B.

Forms, tags, labels, and records

Formally, forms, tags, labels, and records are considered fourth-tier documentation. However, in practice, more often they are incorporated as part of either the procedures or the instructions. Essentially, this tier of documentation provides a master of all the forms, tags, labels, and other records that are used in the organization and either support the procedures and/or are part of the instructions. Some examples of this type of procedures, instructions, and forms are shown in Appendix B.

Records are the documents that support either the quality manual, procedures, and/or instructions. They are less formal, however; they are necessitated by the standards themselves. Figure 5-1 shows the relationship between quality records and the individual elements that require them.

AUDITING

When we think about an audit, most of us may define it as a systematic investigation of the intent, implementation, and effectiveness of selected aspects of the quality systems of an organization or department. By defining it this way, we are identifying that it is not only the system that is to be the subject of the audit, but also the audit process itself that must be adequately documented and controlled against established criteria. The reader is cautioned here to recognize that sometimes, within a company, it can be difficult for the quality manager and/or auditor to gain acceptance for the audit process from individual department managers. They may not treat the subject seriously, since production in most cases is considered far more important than evaluating the effectiveness of their quality system on a periodic basis.

When the audit procedure is defined within a company's quality manual, it is, in effect, endorsed by the senior management as an integral part of the company's operating procedures. In this way, the audit can be given the priority required to allow it to be effective. The method for reporting the results of an audit also requires an established format. This is necessary

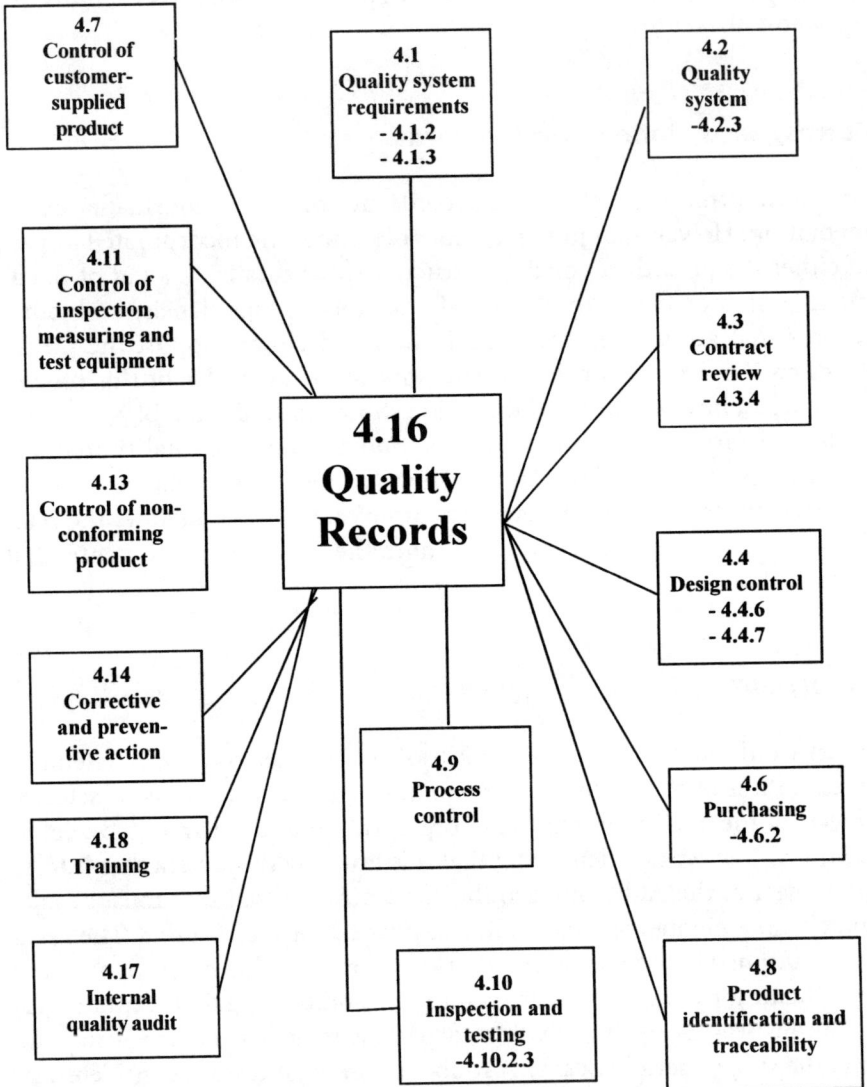

Figure 5-1. Clauses of ISO 9001 that reference clause 4.16 (Control of quality records).

to ensure that the recommendations can be reviewed to determine that the auditor has come to reasonable conclusions in the light of objective evidence that the auditor has identified.

The audit purpose, then, is to collect objective evidence and to permit an informed judgment about the status of the quality system. A typical audit is shown in Figure 5-2. Of course, this applies whether the audit is first, second, or third party. Objective evidence may be evidence that

exists

is uninfluenced by emotion or prejudice

is based on observation

may be stated or documented

is about quality systems

may be quantitative

can be verified

There is also another definition provided by ISO 8402, which states: "An audit is a *systematic* and *independent* examination to determine whether quality activities and related results *comply* with planned arrangements and whether these arrangements are *implemented effectively* and are suitable to achieve objectives" (italics added) (ANSI/ASQC, 1994a). For those who are practicing auditors or who aspire to become one, it is strongly recommended to become acquainted with ISO 10011, which codifies the good practices of auditing.

Based on the official ISO definition, the audit must be systematic. This means that time must be spent planning if maximum benefit is to be gained from available resources. It also must be independent. Whoever carries out the audit must be independent of the activity being audited. With internal audits, the management of a company will choose whomever they consider independent. It will be up to the external auditor to find objective evidence of a failure of independence.

The status of the quality system can be viewed from three perspectives:

1. Do the quality activities comply with planned arrangements? Is the quality system adequately planned to address all the requirements that must be achieved? Is there a clearly defined approach to the requirements?

| STEP 1 | **Prerequisites for an audit**

- *Determine who is going to be an auditor*
- *Determine the standard for auditing*
- *Determine the process knowledge for auditing*
- *Train the team* | **STAGE 1** |

| STEP 2 | **Preparing for the audit**

- *Selection of the audit team*
- *Evaluation of the documentation*
- *Formulation of the audit program*
- *Generating a checklist* | **STAGE 2** |

| STEP 3 | **Performing the audit**

- *Opening meeting*
- *Interviewing*
- *Actual audit: Examining documents, work in progress, workplaces, etc.*
- *Discussion with the team members*
- *Closing meeting* | |

| STEP 4 | **Reporting on the audit**

- *Recording noncompliances*
- *Evaluation of the results*
- *Request for corrective action*
- *Distribution of the report* | **STAGE 3** |

| STEP 5 | **Follow-Up**

- *Verification of corrective action*
- *Re-audit if required* | |

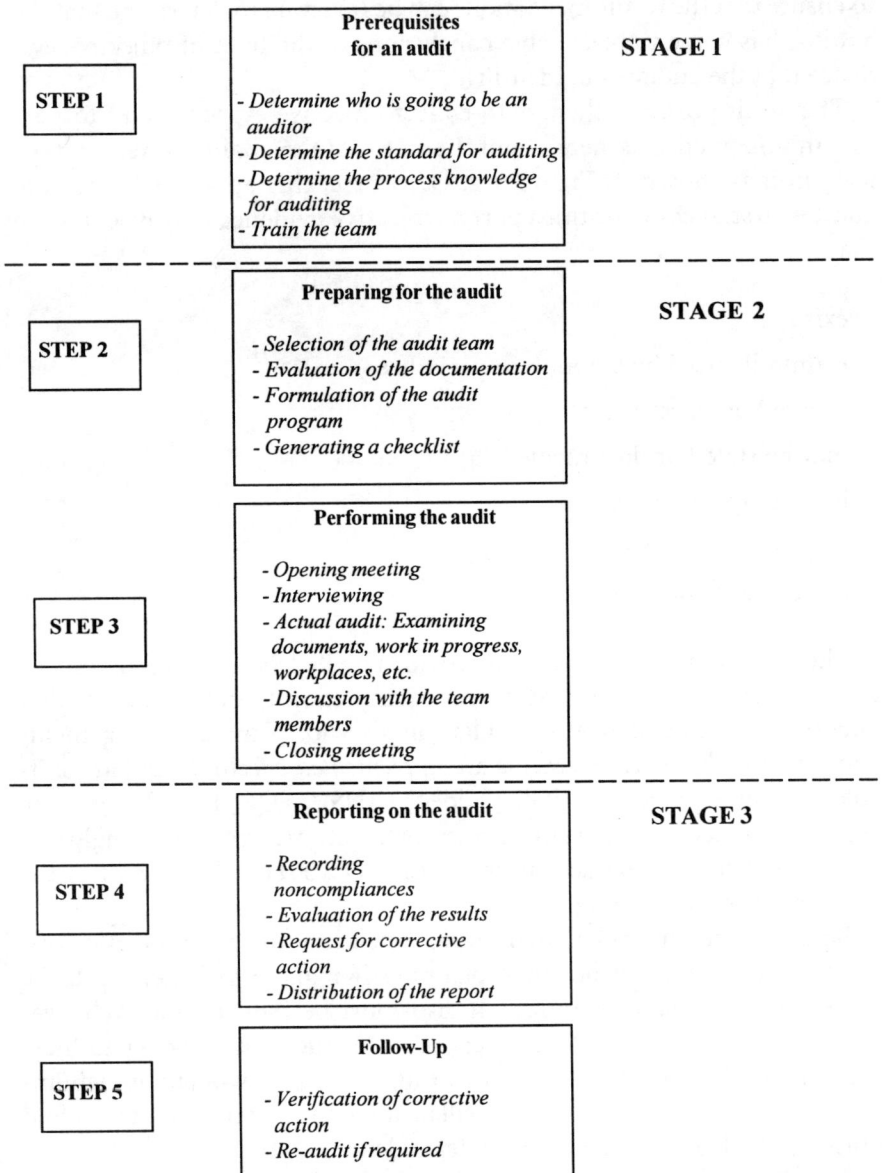

Figure 5-2. Typical audit process.

2. Have these planned arrangements been implemented? Is this approach adequately deployed? Do the processes conform to the defined approach?
3. Are the arrangements implemented effectively? Is the deployment of the approach achieving the requirements?

In seeking objective evidence of each of these aspects, the auditor must not be influenced by preconceived ideas or personal biases.

Types of audits

In auditing, there are three general types of assessments: first, second, and third party. However, within the three types of assessments, there are various categories of audits extending the entire spectrum of the quality function. They include product audit, compliance audit, system audit, gap audit, audit of performance against company objectives, and audit of policies and objectives, to name a few.

All audits are effectively an evaluation of a company's quality management capability to determine:

- Does a system exist?
- Is it implemented?
- Is it effective?

In the final analysis, then, an audit is a "show me" exercise, a dynamic investigation of a company's system that physically moves through the organization and observes the actions of the staff operating their company's policies and procedures.

First-party audits—us on us

The first-party (us on us) audits may be undertaken for a variety of reasons. One very powerful reason is because the quality assurance standard to which the company is working demands that internal audits be performed (clause 4.17.) It is of paramount importance to recognize that organizations seeking ISO 9001 certification have a mandatory requirement to conduct internal audits.

The results of an internal audit will enlighten management to failures so corrective action can be attempted before the failures are found by the external auditor. By the same token, internal audits will also identify good practices to be used in the rest of the organization.

In the final analysis, an internal audit is a tool that management may use in the effective implementation and improvement of the company's quality system.

Second-party audits—us on our suppliers

ISO 9001 does not require a company to audit all of its suppliers, although the wording of the standard has been taken to imply the need. In any case, second-party audits stimulate the supplier to think about their quality system. The independence of an external auditor will often see through the fogging that sometimes occurs, even with companies with the best of intentions. These two outcomes can often result in the supplier taking action to improve the system.

Furthermore, the questions asked by the second-party auditor will often reveal what is important to the purchasing organization. If the supplying organization takes notice, it can concentrate its effort so it will be most effective.

Third-party audits—external assessor on us

Third-party audits are independent of the organization or the supplier; they usually are performed through a certification body. The third-party audit was developed with a number of objectives:

1. Reduce the number of quality management standards being used to audit suppliers.
2. Reduce the number and extent of second-party audits.
3. For those companies registered by a reputable certification body, they will have a distinct marketing advantage over companies yet to achieve registration.

Benefits of auditing

Auditing the system, by properly trained and competent auditors, has the following benefits:

1. The system can be tested for compliance, conformance, and effectiveness. The results of the audit can be used by committed management to implement appropriate changes.
2. The auditor can spot opportunities to further improve an adequate system.
3. The audit by a second-party auditor gives the purchasing organization an opportunity to develop first-line defense mechanisms when it knows its suppliers have weaknesses.

A typical audit

All audits go through a three-stage evaluation. Certainly, all companies have requirements to be met. These requirements are company objectives and policies, contract requirements, and requirements for management of quality.

First stage

In the first stage of the audit, a comparison is made to examine whether the auditee has thought through and defined the means to achieve the requirements. For example, the defined methods will be manifested as a quality manual, written and formalized procedures, work instructions, forms, and computer-driven methods.

This stage has basically two aims. The first is to gain some prior knowledge of the extent to which the auditee has defined its quality system. The second aim is to allow the auditor to consider how the auditee functions and develop an audit schedule and checklist.

Second stage

The second stage compares the defined means of achieving the requirements with the activity. The objective is to confirm the extent to which the methods are being used. The objective evidence is sought by interviewing, observing activities and records, and other means.

In this stage, it may be useful to consider the requirements of ISO 9001 in four distinct areas: major systems requirements, monitors, functional discipline, and key monitors.

Major systems requirements are requirements that must be present in all management systems regardless of the business of the company. They are:

Clause 4.1: Management responsibility

Clause 4.2: Quality system

Clause 4.3: Contract review

Clause 4.5: Document and data control

Clause 4.12: Inspection and test status

Clause 4.13: Control of nonconforming product

Clause 4.14: Corrective and preventive action

Clause 4.15: Handling, storage, packaging, preservation, and delivery

Clause 4.16: Control of quality records

Clause 4.18: Training

Monitor requirements are associated with providing evidence of quality being achieved as defined by the standard itself, or not achieved. These are also the requirements that provide the basis for analysis of information required in the third stage. They are:

Clause 4.7: Control of customer-supplied product

Clause 4.8: Product identification and traceability

Clause 4.10: Inspection and testing

Clause 4.11: Control of inspection, measuring, and test equipment

Clause 4.20: Statistical techniques

Functional discipline requirements deal with a specific activity, often a specific department in a company. They are:

Clause 4.4: Design control

Clause 4.6: Purchasing

Clause 4.9: Process control

Clause 4.19: Servicing

Key monitor requirements demand that management state their intentions and put those intentions into practice. They are:

Clause 4.1.3: Management review

Clause 4.17: Internal quality audit

Third stage

The third stage compares the results of the activity with the requirements. The aim of the third stage is to determine if the overall system is effective in achieving the primary objectives of the company, contact requirements, and the requirements of the standard for quality management systems. This is one of the most difficult parts of the audit analysis. The information gathered during Stage 2 will be used to determine the overall system effectiveness; as a consequence, the results (passing or failing) will be declared. A pictorial view of the three stages is shown in Figure 5-2. The figure also identifies each of the individual steps within each of the stages.

An audit based on the TE supplement

The audit for the TE Supplement is really not any different from any other quality system audit. However, the one major difference is that it follows a very structured approach based on a prescriptive methodology given in the *Tooling and Equipment Quality System Assessment: QSA-TE* (Chrysler, Ford, and General Motors, 1996b). Even though the approach is very descriptive, the fact remains that the QSA-TE seeks the answer to the question: Is the company doing what it says it does? If not, the audit offers opportunities for the company to make improvements. The audit can also

measure the effectiveness of systems and personnel

ensure compliance with quality policies

optimize relationships

identify weaknesses

contribute to technology sharing

reduce customer complaints

To pass a TE Supplement audit, there are two alternatives. The first one has a pass/fail designation. The pass designation is ensured if and when there is no major or minor nonconformance. On the other hand, a fail designation is given if and when more than one major nonconformance exists. In case there is a major nonconformance and/or multiple minor nonconformances, then the status "open" is used. This can be converted to a pass designation within 90 days or other accepted criteria with documented evidence of conformance.

CERTIFICATION

As of this writing, there is no third-party certification requirement for the TE Supplement. A tooling and equipment company, depending on specific customer request, may pursue certification under the ISO 9000 standards. (A typical certification process is shown in Figure 5-3.) However, the guidelines given in the QSA-TE may be used for either self-assessment or as a second-party audit. It must be emphasized, as of this writing no supplier in the tooling and equipment industry may pursue QS-9000 certification.

Before considering certification, any company must consider the following:

1. The first step is to implement a quality system that meets and/or exceeds the QS-9000 requirements.
2. To qualify for certification, it is not enough just to comply with the standard. A document must be created that stipulates a facility's quality-related policies, procedures, and practices. This document is known as the quality manual, and it plays a vital role in the certification process.
3. The facility's quality system must be in operation for a minimum of 3–6 months so that employees are familiar with the system and an evidentiary trail of documents has been created for auditors to review.

After these preliminary steps are taken, a relationship must be established with an accredited registrar. Once the relationship is established, a formal application must be filed. When the application is approved, then the desk audit takes place, followed by the visit to the actual facility. During the audit, the auditor interviews, reviews records, performs detail inspections, observes, and probes into the quality system of the company to document objective evidence.

When the physical audit finishes, the auditor reports the findings and issues the certificate if no major nonconformances are found. A major

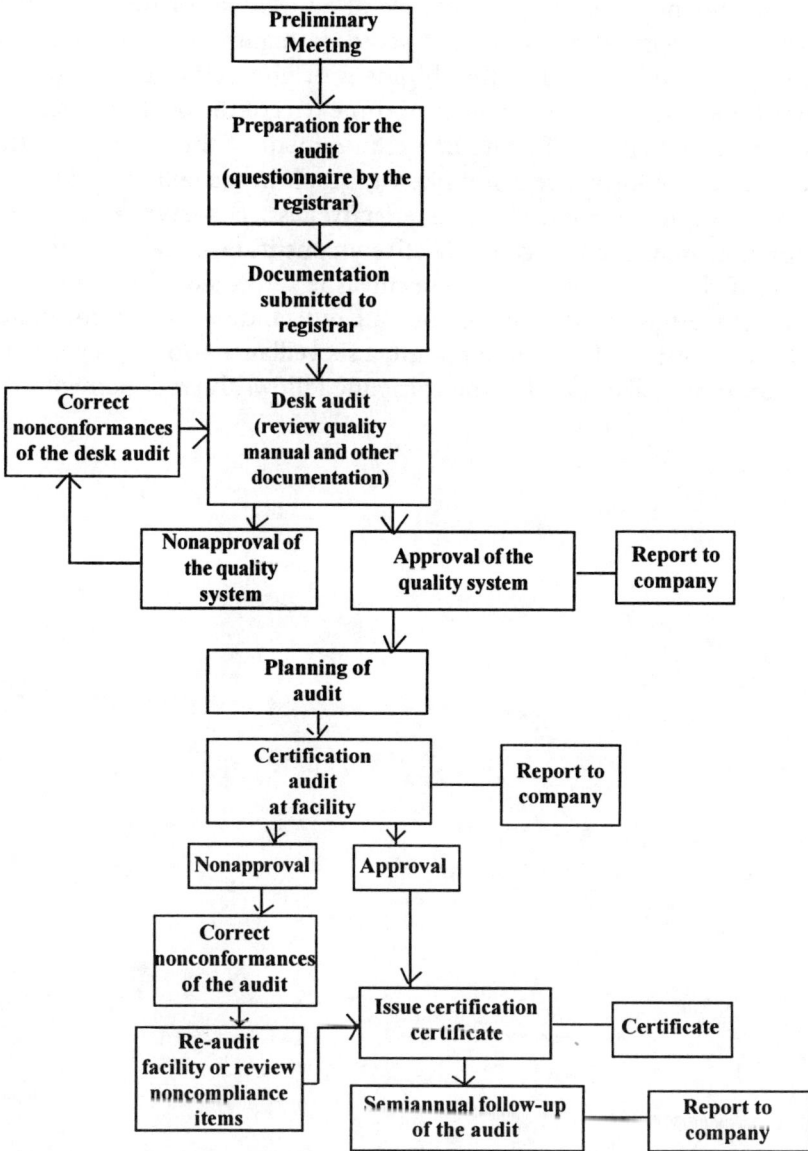

Figure 5-3. A typical third-party certification process.

nonconformance indicates the absence of an element or the total break-down of a system needed to meet a standard requirement. A major non-conformance could result in the shipment of nonconforming product. A minor nonconformance is a failure of some part of the quality system or a single observed lapse following one requirement of the system. In either case, the nonconformance is not likely to result in the failure of the quality system or materially reduce its effectiveness. However, a number of minor nonconformances can collectively constitute a major nonconformance. If there are problems, these must be remedied within a set time frame. Once the registrar closes out all outstanding nonconformances, certification is issued. From that point, a surveillance two times per year is mandatory to make sure that the company follows the requirements.

6

Advanced product quality planning

Advanced Product Quality Planning (APQP) is a discipline that uses many tools to ensure that product quality meets the requirements of the customer. In this chapter, our emphasis is on describing the topic of APQP from a broad perspective rather than identifying and explaining each of the tools used in conducting and implementing APQP. In addition, a special summary of both Chrysler's and Ford's approaches to APQP is discussed. The approach of General Motors (GM) is the default approach identified in the reference manual, *Advanced Product Quality Planning and Control Plan* (Chrysler, Ford, and General Motors, 1995a), distributed by the Automotive Industry Action Group (AIAG).

INTRODUCTION

The goal of both QS-9000 (Chrysler, Ford, and General Motors, 1995d) and the TE Supplement (Chrysler, Ford, and General Motors, 1996a) is the creation and use of quality systems that provide continuous improvement. The emphasis of both quality systems is on defect prevention and the reduction of variation and waste in the supply chain. Both QS-9000 and the TE Supplement have ISO 9001 as a core, augmented by many requirements specific to the automotive industry. Specifically, ISO 9001, clause 4.2.3 indicates the need for quality planning. QS-9000 adds this specification, stating: "[T]he supplier shall utilize the *Advanced Product Quality Planning and Control Plan* (APQP) reference manual."

APQP provides a structured method of defining and executing the steps needed to ensure that products and services satisfy the customer. The goal is to facilitate communication with all involved personnel and ensure that a comprehensive assessment of work practices, tools, and analytical techniques takes place. The result of this goal is to end with a comprehensive plan that ensures that all required steps are completed on time. At least in the automotive world, this process must become part of the everyday life of a company's operations, not only because it is a customer requirement, but, above all, because it is a systematic approach to plan your business.

Just as in any program, there are many individual tools that can make the program work. Therefore, in this chapter, rather than explaining individual tools and techniques, we cover the topic of APQP from a conceptual perspective. To do that, a simple five-step approach is used. Each step consists of inputs, processes, and outputs. The outputs of each phase become the inputs of the next phase.

Step 1. Plan and define the program. This involves the definition of the project goals, using measurable targets that will allow the best possible product, on time, and at an acceptable cost.

- Inputs
  customer requirements and expectations
  internal considerations
  compliance issues
  key product characteristics
  benchmarking results
  goal setting
- Process
  design input via the actions of cross-functional teams
- Outputs
  a project plan
  project goals
  definition of the intended product

Step 2. Product design and development. This yields a definition of the product sufficient to allow development of a manufacturing process and demonstration of stated program goals.

- Inputs
  a project plan
  project goals
  definition of the intended product

- Processes

  Failure Mode and Effect Analysis (FMEA)
  design reviews
  prototypes
  design verification

- Outputs

  drawings
  specifications
  performance expectations

Step 3. Process design and development. This defines the process sufficient to begin production.

- Inputs

  drawings
  specifications
  performance expectations

- Processes

  prototypes
  process reviews and capabilities
  tooling and gaging analysis

- Output

  preproduction control plan

Step 4. Product and process validation. This analyzes the product and process design to demonstrate that it meets program goals.

- Input

  preproduction control plan

- Processes

  process capability
  production validation
  packaging evaluations

- Output

  production control plan

Step 5. Feedback, assessment, and corrective action. This evaluates the product and process so they consistently meet APQP program goals while increasing quality and customer satisfaction.

- Input

  production control plan

- Processes

  variation reduction
  supplier development
  cost reductions

- Outputs

  on-time delivery
  high-grade quality product
  customer satisfaction

THE PROCESS OF APQP

In each of the five steps identified in the introduction of the APQP manual, there are certain activities that are expected. In this section, some of these activities are identified on a per step basis.

Step 1. Plan and define the program. The company must create and document appropriate and applicable documentation for each process. Table 6-1 displays typical documentation for this step.

Step 2. Product design and development. In this step, the company must develop product design features and characteristics to near-final form. Table 6-2 displays typical documentation for this step.

Step 3. Process design and development. The company is responsible for the development of manufacturing systems and related control plans. Table 6-3 displays typical documentation for this step.

Step 4. Product and process validation. The company must have a program for validation of the design, product, and process via a production trial run. Table 6-4 displays typical documentation for this step.

Step 5. Feedback, assessment, and corrective action. In this step, a company must demonstrate an assessment of effectiveness, reduction of variation, and corrective and preventive action. Table 6-5 displays typical documentation for this step.

Table 6-1. Typical Step 1 Documentation

Inputs	Process	Outputs
ɔice of the customer	Integrated effort of	Design goals
• quality function	cross-functional teams	Reliability and quality goals
deployment (QFD)	Effective use of	Preliminary and quality goals
• market research	company-level data	Preliminary process flow chart
• historical warranty	QFD	Preliminary listing of special product
• historical quality	Comparison of similar	and process characteristics
info	FMEAs	Product assurance plan
• team experience	A documented project	Management support
ᴊsiness plan and/or	plan	
marketing strategy		
ᴏduct/process		
benchmark data		
ᴏduct/process		
ᴀssumptions		
ᴏduct reliability		
studies		
ᴊstomer input		

Benefits of APQP

An organization using APQP will benefit in many ways, including the following:

- A reduction in the complexity of product quality planning for the customers and suppliers
- A means for suppliers to easily communicate product quality planning requirements to subcontractors
- A documented and systematic approach of meeting customer requirements
- A formal approach to planning for quality

Why Use APQP?

In addition to the benefits just described, APQP should be used because

- It is the essential characteristic of defect prevention and continuous improvement.
- The auto industry, especially Ford, is placing emphasis on APQP through their *APQP Status Reporting Process* (Ford, 1995).

Table 6-2. Typical Step 2 Documentation

Input	Process	Outputs
Design goals Reliability and quality goals Preliminary and quality goals Preliminary process flow chart Preliminary listing of special product and process characteristics Product assurance plan Management support	Design FMEA Design reviews Prototype build–control plan Team feasibility commitment and management support	Design responsibility activity • Design FMEA • Design for manufacturability and assembly (plan) • Design verification • Design reviews (design verification plan and report) • Prototype build–control plan • Engineering drawings (including math data) • Engineering specifications • Material specifications • Drawing and specification changes From the quality planning team • New equipment, tooling, and facilities requirements • Special product and process characteristics • Gages and testing requirements • Team feasibility commitment and management support

• Chrysler is also placing an added emphasis through their *Product Assurance Planning* manual (Chrysler, 1995).
• General Motors expects you to plan for and institute quality management early in product development. However, GM does not require any special approach other than the standardized approach specified in the APQP reference manual.
• Suppliers are expected to demonstrate the ability to participate in early design activities from concept through prototype and production.
• Quality planning is to be initiated as close as possible to print release. Planning for quality is needed particularly when a company's management establishes a policy of *prevention* as opposed to *detection*.

Table 6-3. Typical Step 3 Documentation

Inputs	Process	Outputs
Design FMEA	Cross-functional team	Packaging standards
Design for	review of the quality	Product/process quality review
manufacturability and	system (manual)	(with the aid of a checklist)
assembly (plan)	Process flow chart	Process flow chart
Design verification	Process FMEA	Floor plan layout
Design reviews (design	Prelaunch control plan	Characteristic matrix
verification plan and	Process instructions	Process FMEA
report)	Preprocess capability	Prelaunch control plan
Prototype build–control	study plan	Process instructions
plan	R&M Guidelines	Measurement systems analysis
Engineering drawings	(Society of Automotive	plan
(including math data)	Engineers, 1993)	Preliminary process capability
Engineering	for tooling and	study plan
specifications	equipment specifically;	Packaging specifications
Material specifications	QS-9000 requires	Management support required
Drawing and	production part	
specification changes	approval plan [PPAP],	
New equipment,	which does not apply	
tooling, and facilities	to TE companies. (This	
requirements	is the use for the	
Special product and	remaining steps as	
process characteristics	well.)	
Gages and testing		
requirements		
Team feasibility		
commitment and		
management support		

- When you use APQP, you provide a systematic plan for the organiza-tion and resources needed to accomplish the quality improvement task.
- Early planning prevents waste (scrap, rework, and repair), identifies required engineering changes, improves timing for new product intro-duction, and lowers costs. The quality planning process is shown in Figure 6-1. The figure shows that the earlier the planning occurs, the more the benefits increase and the better the satisfaction to the customer.

Analytical techniques

An organization using APQP will use many techniques. Some of these techniques are generic to practically every company, and some are very

Table 6-4. Typical Step 4 Documentation

Input	Process	Output
Packaging standards	Production trial run	Production trial run
Product/process quality review (with the aid of a checklist)	Preliminary process capability study	Measurement system evaluation
		Preliminary process capability study
Process flow chart	Production validation testing	Production validation testing
Floor plan layout		Packaging evaluation
Characteristic matrix	Production control plan	Production control plan
Process FMEA	Sign-off	Quality planning sign-off and management support
Prelaunch control plan		
Process instructions		
Measurement systems analysis plan		
Preliminary process capability study plan		
Packaging specifications		
Management support required		

unique, depending on the company and the product. In this section, we identify some of the most common techniques used in the process of APQP. They are the following:

assembly build variation analysis

benchmarking

cause-and-effect diagram

characteristic matrix

critical path method

design of experiments (DOE)

design for manufacturability and assembly

design verification plan and report (DVP&R)

dimensional (or dynamic) control plan (DCP)

mistake-proofing (Poke-Yoke)

process flow-charting

quality function deployment

system and process FMEA

Table 6-5. Typical Step 5 Documentation

Input	Process	Output

Production trial run
Measurement system
  evaluation
Preliminary process
  capability study
R&M Guidelines
Production validation
  testing
Packaging evaluation
Production control plan
Quality planning sign-off
  and management
  support
	Proactive use of the
	  quality management
	  system
	Regular update of the
	  control plans
		Reduced variation
		• SPC: Variable and attribute
		  control charts
		• Special characteristics must
		  meet the indices specified by
		  the customer
		Customer satisfaction
		Delivery and service

Detailed information on these tools is readily available from numerous publishers, training organizations, and software suppliers.

Control plans

A control plan is a written description of the monitoring and control methods for parts and processes. A control plan is usually written with the following in mind:

suggested format depends on the discretion of the supplier

groups of similar parts

use and update throughout the product life cycle

development by a multidisciplinary team

use of the AIAG format if no special request is given by the customer or if the supplier is at a loss for a specific format

Because the control plan is a road map for quality in a given process, there are some specific items that should be included:

1. Product characteristics. Other supplier-identified characteristics may also be included.
2. Product characteristics are variables that have a cause-and-effect relationship with the identified product specifications. Identify those for which variation must be controlled to minimize product variation.

EARLY
EFFECTIVENESS
DUE TO   P
P   R
L   O   C
A   D   U
N   U   S
N   C   T
I   T   O
N   I   M
G   O   E
                                          N   R

→

T I M E L I N E

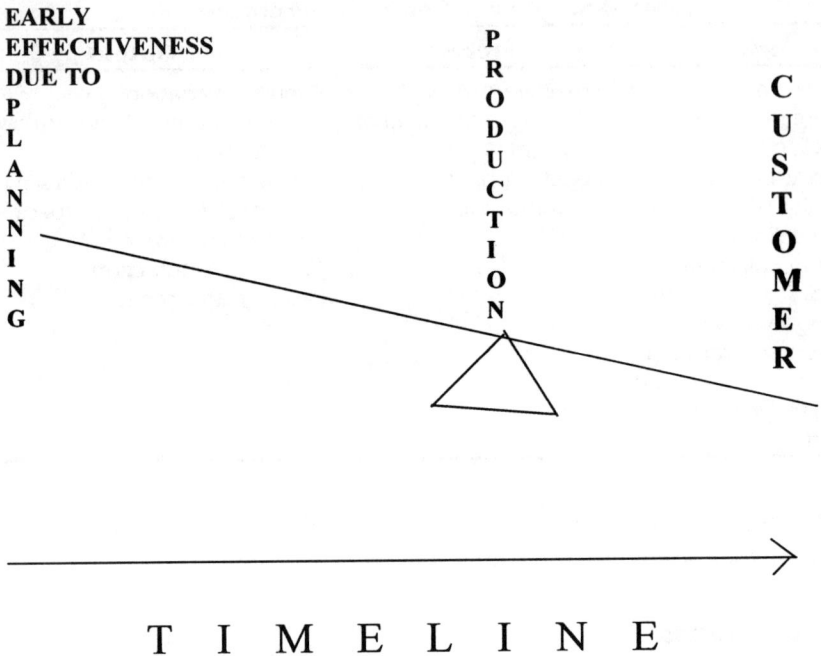

Figure 6-1. A pictorial view of quality planning. Note that as the planning activities get closer to the customer, not only do they become inefficient, but also expensive.

3. Control method is a description of how the operation will be controlled. Use Statistical Process Control (SPC), inspections, mistake-proofing, and sampling plans. Update as change occurs.

4. Reaction plans specify the corrective actions necessary to avoid producing nonconforming products or operating out of control. Identify responsibility.

CHRYSLER'S, FORD'S, AND GENERAL MOTORS' APPROACHES TO APQP

Chrysler, Ford, and General Motors have agreed to a common requirement for APQP. However, both Chrysler and Ford have their own requirements for both content and reporting. The reader is advised to see Chrysler's and Ford's manuals for detailed information. In this section, we only summarize the requirements and guidelines.

Chrysler's approach

Chrysler has agreed that the AIAG version of APQP is the official method-ology and approach of conducting the APQP discipline. However, Chrysler requires the supplier to provide more focus on the product cre-ation process. To facilitate that focus, Chrysler has published *Product Assurance Planning* (1995), in which the framework for the product cre-ation process is explained. Essentially, it covers the necessary activities, responsibilities, deliverables, and measurements required to ensure that products and services will satisfy the customer and meet Chrysler's corpo-rate objectives. In other words, it is a common approach for integrating the areas of quality, reliability, manufacturability, and serviceability into each platform's product development process. The objectives of Chrysler's approach are to continue to build on best practices and improvements identified by platform teams, Extended Enterprise, and worldwide markets and to facilitate feedback from users to the platform quality and reliabili-ty planning organization.

Chrysler's way of approaching the APQP is via the product assurance plan, product assurance team, and major milestone definitions.

Product assurance plans

Product assurance plans are documentation of the Product Assessment Planning (PAP) process. They include specific tasks, deliverables, quality and reliability methods, and tools and measurements that must occur prior to a specifically identified timing milestone in the product develop-ment process to ensure that the product meets customer expectations and corporate objectives. The process is satisfied through the following three levels of PAP:

1. The vehicle-level product assurance plan. This defines vehicle-level objectives, tasks, deliverables, and major program timing milestones to manage effectively the quality, reliability, manufacturability, and serviceability of the vehicle being developed.
2. System-level product assurance plans. These are derived from the vehicle-level plan based on the critical or high-risk product or man-ufacturing systems identified.
3. Component-level product assurance plans. These are derived from system-level PAPs based on critical or high-risk components identified.

Product assurance teams

Product assurance planning is a collaborative activity. As a consequence, teams at several levels are important to facilitate the program at hand. The basic teams required are the following three:

1. Vehicle-level product assurance teams. These are cross-functional teams consisting of general managers, executives, and managers of each vehicle program.
2. System-level product assurance teams. These are cross-functional teams consisting of executives, managers, and product/process personnel who manage the product or process system.
3. Component-level product assurance teams. These are cross-functional teams consisting of managers, product and process personnel, suppliers, and others who manage individual components.

Major milestone definitions

Milestones are major timing events within the product development process that provide points of review to evaluate risk in the program. The major milestones for Chrysler are the following items:

- Concept approval. This is an assessment of program scope, goals, and resource requirements to ensure that they are consistent with the corporate strategy.
- Program approval. This is an assessment by senior management of the detailed objectives, goals, and activities required to complete the program.
- Start of the first program vehicle (F1) build. Program vehicles are built to production design intent to evaluate manufacturing and assembly production process intent for components, systems, and vehicles.
- Start of the first pilot (PO) build. PO pilot vehicles are complete running vehicles built to plant scroll with parts provided from hard tooling.
- Start of the first launch (V1) build. Launch is the start of volume production and assembly of vehicles. V1 must result in complete, quality vehicles that meet all customer requirements and corporate objectives.

Three other significant builds occur during the product development process, and they all occur during the launch phase:

- System tryout (STO) units. These are the first framed body-in-white (BIW) of a new vehicle build with parts approved off production tools.
- C1 pilot vehicles. These are complete running vehicles built by manufacturing at the assembly plants to verify corporate and supplier production tooling, equipment, and processes and to confirm successful resolution of any production tooled product design changes incorporated since the PO build.
- Prevolume production (PVP) vehicles. This marks the beginning of the building of complete, sellable production vehicles.

These three categories are indeed prevention-oriented strategies, with each phase named for the activity for which it is being prepared. Furthermore, within each category there are major elements that make up the PAP. These are

1. Tasks. They identify the activities that must be performed to achieve the objectives.
2. Deliverables. They are defined as the outcomes that should be achieved within a specific phase of the program.
3. Quality and reliability methods and tools. The methods and tools should be based on experience, knowledge, and/or "best practice." The idea here is to select the method and/or tool to accomplish the required task.
4. Measurements. These are used to assess the achievement of a task or deliverable.

Finally, Chrysler provides explicit instructions for the specific methods and tools used in each category in their *Product Assurance Planning* manual publication (1995).

Ford's approach

Just as Chrysler has agreed to the formal AIAG APQP reference manual, so has the Ford Motor Company. However, Ford requires the supplier to provide an official APQP status reporting (Ford, 1995). Essentially, this reporting guideline is to

reinforce the APQP reference manual

develop an evaluation process, process matrices, and status report to assess APQP based on best practices link the APQP evaluation process to Ford's product development process

define the roles and responsibilities for the APQP process

develop a common APQP process for both internal and external suppliers

The process of the status reporting is defined through 13 steps. Steps 1–2 are customer oriented; Steps 3–7 and 9 are oriented to the APQP supplier team; and Steps 8 and 10–13 are oriented to the APQP supplier team leader.

Step 1. Perform risk assessment. The focus here is to identify the supplier and/or product risk. The risk assessment is done to help the extent to which the APQP status reporting process must be performed by the supplier.

Step 2. Communicate APQP status reporting requirements to the supplier. Once the risk assessment is complete, the customer must inform the supplier of the requirement to report status.

Step 3. Form an APQP team. The supplier must form a team, and the team must choose a team leader. The leader is the person responsible for ensuring the APQP status report is complete. Obviously, the team must be cross-functional and multidisciplinary, as well as knowledgeable of the product for which the APQP is to be conducted.

Step 4. Complete header block. Once the team has been formed, they must complete the header block of the actual status report, which has 23 fields. The form is reproduced in Figure 6-2 courtesy of Ford. (The Ford documentation is quite explicit in the directions on how to fill each of the fields.)

Step 5. Determine adjustments. Customer and supplier discuss reporting requirements for the 23 APQP elements. (This is an area in which negotiation is possible. However, the customer must agree to any deviation from the original requirement.) An element is a specific document, task, and/or discipline that must be completed to support the customer's program. The focus of each element is to lay the foundation for program success. The 23 elements are (1) sourcing decision; (2) customer input requirements; (3) design FMEA; (4) design review(s); (5) design verification plan; (6) subcontractor APQP status; (7) facilities, tools, and gages; (8) prototype build control plan; (9) prototype builds; (10) drawings and specifications; (11) team feasibility commitment; (12) manufacturing process flow chart; (13) process FMEA; (14) measurement systems evaluation; (15) prelaunch

Ford Advanced Product Quality Planning Status Report

Date:
Review No.:
Diamond Point:

Supplier	
Location	
Supplier Code	
Risk Assessment	
New: Site ☐ Technology ☐ Process ☐	
Other Risks _____	

Program	
Model Year	
Part No.	
Part Name	
Notice Level	
User Plant(s)	

Team Members	Company/Title	Phone/Fax

Build Level	Material Required Data	Quantity	Concurred		P.I.S.T. %	P.I.P.C. %
			No. SCs	No. CCs		

APQP Elements	GYR Status	Focus Element Rating	Program Need Date	Supplier Timing Date	Closed Date	Resp. Engineer Initials	Remarks or Assistance Required
1) Sourcing Decision							
2) Customer Input Requirements							
3) Design FMEA							
4) Design Review(s)							
5) Design Verification Plan							
6) Subcontractor APQP Status							
7) Facilities, Tools, and Gages							
8) Prototype Build Control Plan							
9) Prototype Builds							
10) Drawings and Specifications							
11) Team Feasibility Commitment							
12) Manufacturing Process Flow Chart							
13) Process FMEA							
14) Measurement Systems Evaluation							
15) Pre-Launch Control Plan							
16) Operator Process Instructions							
17) Packaging Specifications							
18) Production Trial Run							
19) Production Control Plan							
20) Preliminary Process Capability Study							
21) Production Validation Testing							
22) Production Part Approval (PSW)							
23) PSW Part Delivery at MRD							

COMMENTS

Figure 6-2. Advanced Product Quality Planning status reporting form. Copyright Ford Motor Company 1995. (Courtesy of Ford Motor Company. Used with permission.)

control plan; (16) operator process instructions; (17) packaging specifications; (18) production trial run; (19) production control plan; (20) preliminary process capability study; (21) production validation testing; (22) Part Submission Warrant (PSW); and (23) PSW part delivery at Material Review Date (MRD).

It must be noted that 19 of these elements are the standard APQP elements in the APQP reference manual. Only four of these are strictly additional requirements for Ford. However, even these four are founded in the industry standards. The four additional requirements are:

1. Sourcing decision
2. Customer input requirements
3. Subcontractor APQP status
4. PSW part delivery at MRD

Step 6. Determine program need dates. This is a schedule issue, and the last possible date the element can be completed without adversely affecting quality, cost, or delivery of the program must be defined. To define the last possible date, a supplier may use the critical path methodology.

Step 7. Communicate timing dates. The APQP supplier team leader should ensure that anyone who will need to input information in the APQP status reporting process understands what they are required to do.

Step 8. Report against timing dates. Once the program need dates and supplier timing dates are determined, the supplier APQP team leader needs to ensure that the supplier APQP team members understand how to report against them.

Step 9. Assess the status report elements. This measures the quality of element completion by rating if all key expectations have been met. For the convenience of the supplier, Ford has developed a checklist—included in their guidelines—to facilitate the process.

Perhaps one of the most misunderstood elements of this process, this step also addresses the GYR status. GYR stands for Green—the element meets expectations; Yellow—a recovery plan is in place; and Red—the element is at risk and no recovery plan exists. The GYR rating communicates the status of element completion by the program need date. To rate a given element with one of the colors, the checklist is used, which has a rating scale from zero to 3.

A zero rating means that less than half of the expectations are completed. Furthermore, it indicates that management action is needed. A zero rating in early review of a program is acceptable.

A 1 rating means that at least half of the expectations are completed. It indicates progress toward the element completion. If the 1 rating is in the early stages of the program, it is acceptable. However, later in the program it indicates a signal for management action.

A 2 rating means that all expectations are met. However, it must be achieved by the program need date for the element to be considered closed.

A 3 rating means that all expectations are completed and one or more boxes are checked in the "exceeds expectations" category.

Step 10. Gather element data from appropriate team members. Once the team members assess the elements for which they are responsible, the team leader needs to collect that data.

Step 11. Enter the GYR rating focus element rating and revised timing dates. The data must be gathered and entered on the APQP status report whether the data are summarized by the team leader or the team.

Step 12. Assign responsible engineers to Yellow and Red elements. If an element is Yellow or Red, the supplier APQP team leader needs to assign an engineer to investigate and resolve or contain the issues.

Step 13. Review completed APQP status report with customer. Meetings should be held periodically to review the APQP status report. As delivery date approaches, meetings may occur once a week.

The benefits of doing this reporting are to avoid crisis management, reduce miscommunication, lower cost, improve quality, and improve on-time delivery.

General Motors' Approach

Whereas General Motors does not have any special approaches to APQP, the reader should not interpret this as a lack of interest on the part of General Motors for APQP. The reader should be familiar with the standardized approach explained in a cursory format at the beginning of this chapter and delineated in detail in the formal reference manual, *Advanced Product Quality Planning and Control Plan* (Chrysler, Ford, and General Motors, 1995a).

7

Reliability and maintainability guidelines

This chapter summarizes the discipline of Reliability and Maintainability (R&M) as used in the tooling and equipment industry. Specifically, we define the discipline and address the five phases of Reliability and Maintainability. In addition, we discuss some common tools.

INTRODUCTION

Reliability and Maintainability are not tools to be used in specific tasks. Rather, R&M is a discipline. It is founded on several techniques that are meant to direct both machine suppliers and users beyond the question, Will it work? to a quantifiable analysis of how long it will work without failure.

To understand R&M, we must understand its components. First, *reliability* is the probability that machinery/equipment can perform continuously, without failure, for a specified interval of time when operating under stated conditions. Second, *maintainability* is a characteristic of design, installation, and operation usually expressed as the probability that a machine can be retained in, or restored to, specified operable condition within a specified interval of time when maintenance is performed in accordance with prescribed procedures.

R&M are the vital characteristics of manufacturing machinery and equipment that enable its users to be "world class" competitors. After all, efficient production planning depends on a process that yields high-grade parts at a specific rate without interruption. What makes R&M worth pursuing is the fact that it allows the manufacturer of a specific equipment or tool to be able to predict a specified quality level. This predictability is the key ingredient in maintaining production efficiency and the effective deployment of just-in-time principles.

As important as R&M is to any organization, in order for it to work there must be a cooperative effort between the supplier and user of manufacturing machinery and equipment. Both must understand which equipment performance data are needed to ensure continued improvement in equipment operation and design and must exchange this information on a regular basis.

R&M, as it is used in the tooling and equipment industry, is organized in five phases: concept, development/design, building and installation, operation and support, and documentation/conversion/transition.

BENEFITS OF RELIABILITY AND MAINTAINABILITY

There are many benefits of implementing a program that includes R&M. Perhaps the most important benefit of R&M is improved availability of the machine and/or the equipment. A strong second benefit is the fact that improved R&M leads to fewer total life cycles necessary to maintain the competitive edge.

Other benefits include

- Supplier benefits

   reduced warranty costs
   reduced build costs
   reduced design costs
   improved user relations
   higher user satisfaction

- User benefits

   unscheduled downtime reduced/eliminated
   reduced maintenance costs
   stabilized work schedule
   improved profitability
   improved just-in-time capability
   lower overall cost of production
   more consistent part/product quality

IMPLEMENTING RELIABILITY AND MAINTAINABILITY THROUGH THE LIFE CYCLE PROCESS

As mentioned above, R&M depends on a cooperative effort between suppliers and users. For that cooperation to be developed and maintained, good communications must exist between the two. This communication must begin at project conception and continue through the entire life of the equipment to ensure that equipment problems will be identified, root causes determined, and corrective action implemented. A typical process for optimizing the implementation of R&M follows the life cycle (see Table 7-1), which consists of five phases.

Phase 1. Concept

Research and development (R&D) are begun in this phase. The result is a proposal. The essence of this phase is that both the user and supplier must work together to establish the requirements.

Phase 2. Development/design

In the development/design phase, the majority of the life cycle cost is determined. In addition, safety, ergonomics, accessibility, and other maintainability issues are designed into the system. Design reviews ensure that the planned design is likely to meet all requirements in the most cost-effective way, considering all variables and constraints and with special attention to maintainability.

At this phase, the design must include suitable test plans, agreed to by both the user and the supplier, to demonstrate compliance to requirements and provision for teardown and reassembly in the user's plant.

Phase 3. Build and install

In the build and install phase, the results of the R&M should be monitored, and issues that could affect R&M must be communicated back to the design. Special attention should be given to the following:

Table 7-1. Basic Project Methodology Following the Life Cycle

Conceptual focus	Design (concept and detail)	Implementation	Operation and support	Transition
Confirm strategic directions, concurrent initiatives, and focus	Analyze and define major functions and projects	Finalize architectures, sourcing, and configuration	Initiate maintenance and support procedures and services	Monitor requirements and performance measures
Identify critical goals, metrics, and objectives	Assess key alternative concepts	Prepare system	Document	Encourage change
Global model of processes and information	Prepare design approach, requirements, constraints, and details	Develop systems, tests, and procedures	Develop contingency procedures	Support system development environment
Organize teams and assess current position	Compare costs and benefits	Integrate and test the system	Train personnel	Skills transfer
Communicate conceptual vision	Communicate selected program plans and specifications	Publish and start installation and monitor results	Publish, support, and monitor results	Publish and support business

Maintenance procedures are developed by the people who are involved in the process.

Appropriate and applicable training starts at this phase and continues throughout the project as needed.

Machine acceptance testing.

R&M database collection begins during machine acceptance testing. All problems encountered during this phase should be documented for future reference and as candidates for continuous improvement.

The machine will be transferred from the manufacturing machinery and equipment supplier's location to the user's plant. Critical assembly processes should be identified during teardown.

Installation is a critical step. The machine has to be reassembled to the build requirements. Special attention should be given to the critical assembly processes identified during teardown.

Some infant mortality failures may be present during initial startup. Every effort should be taken to eliminate infant mortality failures during the installation and debugging period.

Phase 4. Operation and support

In the operation and support phase, the equipment has been delivered and installed at the user's location and is fully operational. Data collection and feedback are very important in this phase.

Phase 5. Documentation/conversion

The documentation/conversion phase is the end of the expected life of the machine. If an increasing failure rate has resulted in increasingly expensive maintenance, the machine may require decommissioning or may be rebuilt to a good-as-new state. Alternatively, the machine may still be in good condition, but production needs may have changed, requiring the machine to go through major conversion to be used for the manufacture of other products. When decommissioning, rebuilding, or conversion occur, the feedback from the user plant should be recorded.

SELECTED TOOLS USED IN RELIABILITY AND MAINTAINABILITY ON A PER PHASE BASIS

There are many tools that one may use in pursuing R&M. Here we address some of the most common ones on a per phase basis.

Phase 1. Concept

Mean time between failures

The mean time between failures (MTBF) occurrences are the sum of the operating time of a machine divided by the total number of failures (Society of Automotive Engineers, 1993).

Mean time to failure

Mean time to failure (MTTF) is a figure-of-merit used to quantify system effectiveness; it depends on system reliability and equals the average time until an unrepairable item fails. It is measured in operating hours and is a basic indicator of reliability for unrepairable items. It is also the reciprocal of the failure rate. MTTF is equal to the total number of life units of an item divided by the number of failures within that population in a particular measurement interval under stated conditions (Omdahl, 1988).

Mean cycles between failures

Mean cycles between failures (MCBF) is the average number of cycles between failure occurrences and is equal to the sum of the operating cycles of a machine divided by the total number of failures (Society of Automotive Engineers, 1993).

Mean time to repair (or to replace)

The mean time to repair (or to replace) (MTTR) is a factor of mean time between maintenance (MTBM); it refers to the mean time between item replacement and is a major parameter in determining spare parts requirements. On many occasions, corrective and preventive maintenance actions are accomplished without generating the requirement for the replacement of a component part. In other instances, item replacements are required, which in turn necessitates the availability of a spare part and an inventory requirement.

In essence, MTBM is a significant factor, applicable in both corrective and preventive maintenance activities involving item replacement, and is a key parameter in determining support requirements. A maintainability objective in a system design is to maximize MTBR when feasible.

Mean cycle to repair

Mean cycle to repair (MCTR) refers to the average cycles needed to restore machinery or equipment to specified conditions (Society of Automotive Engineers, 1993).

Phase 2. Development/design

Reliability analysis

As already mentioned, reliability is the probability that a system or product will perform in a satisfactory manner for a given period of time when used under specified operating conditions. Reliability is an inherent characteristic of design; therefore, it should be considered at the inception of a project and followed through its life cycle.

Reliability may be analyzed using block diagrams, series systems, or parallel systems.

Block diagrams

Block diagrams are planning tools used to recognize how each equipment component contributes to the functional reliability of the equipment. They are used to calculate system reliability based on component reliabilities. They are also used to determine the minimum component reliability necessary to achieve a desired system reliability.

Series systems

Under the series systems configuration, all components must operate successfully for the system to function correctly. The total reliability of such a system is given by the product of each component in the series as

$$R_T = R_1 \times R_2 \times R_3 \times \ldots R_n$$

where $R_1 \ldots R_n$ is the reliability of each of the components.

Parallel systems

Under the parallel systems configuration, a system is considered to have parallel elements if any element can fail without degrading the performance of the system. A system may be considered partially parallel if the failure of a component degrades the system performance to a predictable level but does not prevent the system from functioning. A parallel system is not considered to have failed unless all redundant subsystems have failed. The reliability equation for calculating total parallel system reliability is

$$R_T = 1 - [(1 - R_1) \times (1 - R_2) \times \ldots \times (1 - R_n)]$$

where $R_1 \ldots R_n$ is the reliability of each of the components

Reliability prediction

Reliability prediction is an estimate of the reliability of a product based on design information derived from system architecture, parts information, and system test data using reliability models that are usually obtained from government and industry sources. It is used to decide if preliminary estimates of reliability performance are satisfactory. If the results are acceptable, the design/development process may continue. If not, the design or the requirements—or both—may need to be modified. There are three basic approaches to reliability prediction: comparison, fundamentals, and trends.

Comparison approach

The comparison approach utilizes the techniques that drive predictions based on similarities with existing systems. It is usually used early in the design cycle, before the operating and environmental stresses have been determined.

Fundamentals approach

The fundamentals approach consists of building a composite failure rate based on estimates of individual contributing causes. Modifiers can be used to adjust the failure rate up or down. This approach is used later in the design cycle, after operating and environmental stresses have been determined.

Trends approach

The trends approach consists of time-trend or growth-curve estimates based on published test data or engineering experiences. This approach is used when experience has shown that the reliability of a product has been changing at a certain rate over a period of time. An example of this is the reliability of integrated circuits that have been improving each year for the last several years.

Accelerated life testing

Accelerated life testing may be conducted on critical components according to customer's and/or regulatory requirements.

Failure mode and effect analysis (FMEA)

As we discussed above, the FMEA is a requirement for both QS-9000 and the TE Supplement. Specifically, it is a requirement for both APQP and R&M. Because of its significance, let us look at it in more detail. An FMEA is a strict methodology to evaluate a system, a design, a process, and/or a service for possible ways in which failures can occur. For each failure, which can be a bona fide or a potential failure, an estimate is made of its effect on the total system and of its seriousness. In addition, a review is made of the action being taken (or planned) to minimize the probability of failure or to minimize the effect of failure (Stamatis, 1995a).

This simple, but straightforward, approach can be technical (quantitative) or nontechnical (qualitative), utilizing three main factors for the identification of the specific failure: occurrence (how often the failure occurs), severity (how serious the failure is), and detection (how easy or difficult it is to detect the failure).

The complication of the approach always depends on the complexity of the problem, as defined by the following:

1. *Safety.* Injury is the most serious of all failure effects. In fact, in some cases it is of unquestionable priority, and at this point it must be handled either with a Hazard Analysis (HA) and/or Failure Mode Critical Analysis (FMCA).

2. *Effects on downtime.* How are repairs made? Can repairs be made while the machine is off-duty or while the machine is operating?

3. *Access.*   What hardware items must be removed to reach the failed component? This area will be of great importance as environmental laws are changed to reflect world conditions for disassembly.

4. *Repair planning.*   Repair time, maintainability, repair tools, cost, recommendation(s) for changes in design specifications. Here, mistake-proofing (Poke-Yoke), DOE, or design for manufacturability may be considered.

To carry this methodology to its proper conclusion, the following four prerequisites must be understood:

1. *Not all problems are important.*   This is fundamental to the entire concept of FMEA because, unless internalized, you will spend too much time "chasing fires" in the organization. The fact is that some problems have higher priority than others. FMEA helps identify this priority.

2. *You must know the customer.*   The definition of *customer* normally is thought of as the "end user." However, a customer may also be defined as a subsequent or downstream operation, as well as a service operation. When using the term *customer* from an FMEA perspective, the definition plays a major role in addressing problems. For example, as a general rule in the design FMEA, one views the customer as the end user, while in the process FMEA the customer is viewed as the "next" operation in line. This next operation may indeed be the end user, but it does not have to be. Once you define your customer (internal, intermediate, or external), you may not change the definition—at least not for the problem at hand—unless you recognize that by changing it you may indeed have changed your problem and/or consequences.

3. *You must know the function.*   It is imperative to know the function, purpose, or objective of what you are trying to accomplish, otherwise you are going to waste time and effort in redefining your problem based on situations. If you must, take extra time to make sure you understand the function or purpose of what you are trying to accomplish.

4. *You must be prevention oriented.*   Unless you recognize that continual improvement is in your best interest, the FMEA is going to be a "static" document to satisfy your customer or market requirements. The push for this continual improvement makes the FMEA a "dynamic" document that changes as the design and/or process changes, with the intent *always* to make a better design and/or process.

Why do we do FMEAs?

The propensity of managers and engineers to minimize the risk in a particular design and/or process has forced us to look at reliability engineering not only to minimize, but also to define, the risk. Obviously, the risk is a multifaceted issue; however, from a generic perspective, one may define it based on: management emphasis; market pressures; customer requirements; safety; legal issues; statutory requirements; public liability; development technical risks; warranty and service costs; competition; and so on. These risks can be measured by reliability engineering and/or statistical analyses. However, because of their complexity, the FMEA has extracted the basic principles without the technical mathematics and has provided us with a tool that anybody committed to continual improvement can utilize.

Statistical Process Control (SPC) is another technique that provides the impetus for the implementation of an FMEA, especially for a process FMEA. SPC provides information about the process in regard to changes. These changes are called common and special causes. From an FMEA perspective, we may look at the common causes as failures that are the result of inherent failure mechanisms; as such, they can affect the entire population. This is a cause for examining the design.

On the other hand, special causes are looked at as failures that result from part defects and/or manufacturing problems; as such, they can affect a relatively small population. In this case, there is cause for examining the process.

Customer requisition, of course, strongly influences the reason we may be doing an FMEA. For example, all major automobile companies in their quality systems, as well as in their supplier certification programs, require an FMEA program. Courts—through product liability—may require some substantiation as to what level of reliability your products perform.

International standards such as the ISO 9000 series may define the program of documentation in your design. For example, the Product Liability Directive of EU 1985 (Kolka and Scott, 1992) stipulates that manufacturers of a product will be held liable, regardless of fault or negligence, if a person is harmed or an object is damaged by a faulty or defective product (this includes exporters into the European Union [EU] market). This liability directive essentially reverses the burden of proof of fault from the injured to the producer. Quality systems incorporating specific tools such as FMEA or fault tree analysis (FTA) or FMCA with safety prevention provisions will be particularly important in protecting a company from unfounded liability claims. Furthermore, proposed safety directives would oblige manufacturers to monitor the safety of their products throughout the life of the product.

A partial list of other benefits of the FMEA includes:

- Improving the quality, reliability, and safety of the products
- Improving the company's image and competitiveness
- Helping increase customer satisfaction
- Reducing product development time and costs
- Helping select the optimum system design
- Helping determine the redundancy of the system
- Helping identify diagnostic procedures
- Establishing a priority for design improvement actions
- Helping identify critical and/or significant characteristics
- Helping in the analysis of new manufacturing and/or assembly processs

Even though all the above reasons are well worth the effort of conducting an FMEA, the most important reason for performing an FMEA is the need to improve. Unless this need is part of the culture of the organization, the FMEA program is not going to be successful.

What is the language of the FMEA?

To understand the FMEA, one must understand its language. There are several terms that one must understand.

Causes of failure. The "root cause" of the listed failure. Next to the function, cause of failure is perhaps the most important section of the FMEA and points the way toward preventive and/or corrective action. The more focused you are on the root cause, the more successful you will be in eliminating failures. In this section, be careful not to be too eager for solutions lest you fall victim to short-term remedies rather than the complete elimination of the real problem(s). Examples of terms for design are *wall thickness, vibration, shock load,* and *torque specifications.* Examples of terms for process are *voltage surge, dull tools, improper setup,* and *worn bearings.*

Current control(s). Controls that exist to prevent the cause(s) of the failure from occurring in the process phase. Examples include any of the SPC tools, capability, and operator(s) training.

Design verification. Controls that exist to prevent the cause(s) of the failure from occurring in the design phase. Examples include design guidelines, design reviews, and specific specifications.

Effects of failure. The outcome of the failure mode on the system and/or the product. In essence, the effects of the failure have to do with the question of what happens when a failure occurs. We must understand, however, that the effects of the failure must be addressed from two points of view. In the local view, the failure is isolated and does not affect anything else. In the global view, the failure can and does affect other functions and/or components. It has a domino effect. Generally speaking, the failure with a global effect is more serious than the failure of a local nature. The effect of the failure also defines the severity of a particular failure. For example, a local failure could be a parking light bulb failure, while a global failure would be a power brake failure. In the first case, one can identify the failure as a nuisance; in the second case, a catastrophic failure is imminent.

Failure. The problem; the inability of a component, subsystem, or system to perform to design intent. This inability can be defined as both known and potential. Of great importance here is that, when potential failures in terms of functional defectives are identified, the FMEA is fulfilling its mission of prevention. Functional defectives are failures that do not meet the customer's requirements, but the products are shipped with the failures anyway because the customer will never know the difference, never find out, find out but whatever is delivered can be used, or find out but whatever is delivered has to be used because there are no alternatives. Examples of failure terms are *broken, worn, corrosion,* and *noise.*

Function. The task that a component, subsystem, or system must perform. This function is very important in understanding the entire FMEA process. It has to be communicated in a way that is concise, exact, and easy to understand, using no jargon. To facilitate this, it is recommended that an active verb be found to describe the function. The active verb, by definition, defines performance, and performance is what a function is. Examples of active verbs are *position, support, retain,* and *lubricate.*

Product validation. Controls that exist to prevent the cause(s) of the failure from occurring and to validate repeatability for certain processes.

What are the mechanics of failure mode and effect analysis?

To do an effective FMEA, a team must generate the FMEA. The reason for this is that the FMEA should be a catalyst to stimulate the interchange of ideas between the functions affected. A single engineer or any other single person cannot do it.

The team should be made of five to nine persons (preferably five). All team members must have some knowledge of team behavior, the task at hand, the problem to be discussed and some kind of either direct or indirect ownership of the problem; above all, they must all be willing to contribute. Team members also must have cross-functional and multidisciplinary skills. Furthermore, whenever possible and/or needed, it is best to have the customer and/or the supplier actively participate.

A design FMEA is a systematic method to identify and correct any known or potential failure modes before the first production run. A first production run is viewed as the run by which you produce a product and/or service for a specific customer with the intent of getting paid. This definition is important because it excludes initial sample runs (ISRs), trial runs, sometimes the prototype run(s), and so on. The threshold of the first production run is important because to that point modification and/or change of the design is not a major undertaking. After that point, however, the customer gets involved through the letter of deviation, waiver of change, or some other kind of formal notification.

Once these failures have been identified, we rank them and set priorities for them. The leader (the person responsible for a design FMEA) should be the design engineer, primarily because the leader is most knowledgeable about the design and can best "anticipate" failures. The quality engineer should be designated as the facilitator.

The minimum makeup of the team for the design should be a design engineer and a process (manufacturing) engineer. Anyone else who can contribute or who the design engineer feels is appropriate may also participate. A typical design team also may include a test/development engineer, a reliability engineer, a material engineer, and a field service engineer (customer's voice).

Of great importance in the makeup of the team is that the team must be cross-functional and multidisciplinary. However, remember that there is no such thing as "the" team. Each organization must define their optimum team participation, recognizing that some individuals may indeed hold two or more positions at the same time.

The focus of the design is to minimize failure effects on the system by identifying the key characteristics of the design. These key characteristics may be identified through customer requirements, engineering specifications, industrial standards, government regulations, and courts (through product liability).

The objective of the design FMEA is to maximize the system's quality, reliability, cost, and maintainability. It is important here to recognize that in design we have *only* three possibilities to look at defects:

1. Components. The individual unit of the design.
2. Subsystem or subassembly. Two or more combined components.
3. System or assembly. A combination of components and subsystems for a particular function.

Regardless of the level of the design, the intent is the same—no failures in the system.

To focus on these objectives, the design team must use consensus for their decision; more importantly, they must have the commitment of their management.

The timing of the design FMEA is initiated during the early planning stages of the design and is continually updated as the program develops. As a team, you must do the best you can with what you have, rather than wait until all the information is in. By then, it may be too late.

A process FMEA is a systematic method for identifying and correcting any known or potential failure modes before the first production run. (The first production run was defined in the design section.) Once these failures have been identified, we rank them and assign them priorities.

The leader (the person responsible for the process FMEA) should be the process/manufacturing engineer, primarily because the leader is most knowledgeable about the process structure and can best anticipate failures. The quality engineer should be designated as the facilitator. The minimum makeup of the team for the process is a process/manufacturing engineer, a design engineer, and an operator(s). Anyone else who can contribute or who the process engineer feels is appropriate may also participate. A typical process team also includes: a quality engineer, a reliability engineer, a tooling engineer, and an operator(s).

Of great importance in the makeup of the team is that the team must be cross-functional and multidisciplinary. Remember, there is no such thing as "the" team. Each organization must define their optimum team participation, recognizing that some individuals may indeed hold two or more positions at the same time.

The focus of the process FMEA is to minimize production failure effects on the system by identifying the key variables. These variables are the key characteristics of the design, but now, in the process, they have to be measured, controlled, monitored, and so on. This is the point at which SPC comes alive.

The objective of the process FMEA is to maximize the system quality, reliability, and productivity. The objective is a continuation of the design FMEA—more or less—since the process FMEA assumes the objective to be as designed. Because of this, potential failures that can occur because of a design weakness are not included in a process FMEA. They are only mentioned if those weaknesses affect the process.

The process FMEA does not rely on product design changes to overcome weaknesses in the process, but it does take into consideration a product's design characteristics relative to the planned manufacturing or assembly process to ensure that, to the extent possible, the resultant product meets customer needs, wants, and expectations.

Another important issue in a process FMEA is that it is much more difficult and time consuming than the design. The reason for this is that in a design FMEA we have three possibilities for analysis, and in the process FMEA we have six, and each of these may have even more: personnel, machine, method, material, measurement, and environment.

To show the complexity of each of these possibilities, let us take the machine as an example. Some of the contributing failures may, in fact, be in one of the following categories: tools, workstation, production line, process itself, gages, operator(s), and so on.

Again, just as in the design, the process team must use consensus for their decision but, more importantly, they must have the commitment of their management.

To facilitate this consensus and gain the commitment of management, the team in both design and process FMEAs must set specific improvement goals. To do that, the leader of the team must use the following specific behaviors:

Focus the team on result areas. The leader should stress the importance of goal setting for focusing team efforts and discuss possible result areas the team may choose.

Review existing trends, problem areas, and goals. The leader should share with the team data that could shed light on the team's performance and indicate possible areas for improvement.

Ask (actively solicit) the team to identify possible areas for investigation. The leader should actively—at all times—ask for complete participation from all team members. The more participation there is, the more ideas there will be. Brainstorming, affinity charts, and force field analysis are some of the tools that can be used by the leader to facilitate an open discussion by all.

If possible, ask the team to identify the goal. The leader, after a thorough discussion and perhaps ranking, should have the team agree on the selected goal. This will enhance the decision and will ensure commitment.

Specify the goal. It is the leader's responsibility, once the goal has been set, to identify the appropriate measures, time, and amount of improvement. As a rule, cost is not considered at this stage. The cost is usually addressed as part of another analysis called value engineering.

In the final analysis, regardless of the level at which the FMEA is being performed, the intent is always the same—no failures in the system.

The timing of the process FMEA is initiated during the early planning stages of the process, before machines, tooling, facilities, and so on are purchased. The process FMEA, just like the design FMEA, is continually updated as the process becomes more clearly defined. As for the displaying format, there are many forms available; however, no particular format is the standard for everyone. Typical forms used in design and process FMEAs are shown in Figures 7-1 and 7-2, respectively. Figure 7-3 and Figure 7-4 show typical FMCA forms. An FMEA form that also includes a control plan is shown in Figure 7-5.

FMEA guidelines

The abundant availability and singularity of use in different organizations do not allow specific discussion of FMEA guidelines. Rather, a general discussion follows.

The guidelines are numerical values based on certain statistical distributions that allow us to set priorities for the failures. They are usually of two forms, qualitative and quantitative. In the qualitative form, one bases the meaning on theoretical distributions such as normal, norm-log (skewed to the left), and discrete distributions for the occurrence, severity, and detection, respectively.

Potential Failure Mode and Effect Analysis (Design Form)

System _____
Subsystem _____
Component _____
Model _____

Design responsibility _____
Due Date _____
Core Team _____

FMEA Number _____
Page ____ of ____
Prepared By _____
FMEA date (Orig.) _____ Rev. ____

Item/ Function	Potential Failure Mode	Potential Effect(s) of Failure	S E V	C R I T	Potential Cause(s) Mechanism(s) of Failure	O C C U R	Current Design Controls	D E T E C	R P N	Recommended Action(s)	Responsibility & Target Completion Date	Action Taken	S E V	O C C	D E T	R P N

ACTION RESULTS

Figure 7-1. A typical design FMEA form.

Potential Failure Mode and Effect Analysis (Process Form)

System	Process responsibility _____	FMEA Number _____
Process	Due Date _____	Page ___ of ___
Item	Core Team _____	Prepared By _____
Model _____		FMEA date (Orig.) _____ Rev. _____

Purpose/ Function and/or require-ments	Potential Failure Mode	Potential Effect(s) of Failure	S E V	C R I T	Potential Cause(s) Mechanism(s) of Failure	O C C U R	Current Process Controls	D E T E C	R P N	Recommended Action(s)	Responsibility & Target Completion Date	Action Taken	ACTION RESULTS			
													S E V	O C C	D E T	R P N

Figure 7-2. A typical process FMEA form.

Part Number (1) _____

Assembly Number (2) _____

Responsible engineer (3) _____

Production release date (4) _____

Page (5) _____ of _____

Date (6) _____

Line number (7)	Cross-reference number (8)	Circuit location (9)	Enter the part/ component number/name (10)	Function(s) & specification(s) (11)	Potential failure mode(s) (12)	System effect (0=Unsafe condition) (13)	Unsafe (14)	Cause(s) of failure (15)	Internal or external counter-measures (controls) (16)	Severity (17)	Base failure rate λB (18)	Failure mode ratio (19)	Effectiveness (20)	Risk priority number (RPN) (21)

104

1. Part number. Enter the part number under consideration.

2. Assembly number. Enter the number on the part or drawing or part list.

3. Responsible engineer. Enter the name of the responsible engineer.

4. Production release date. Enter the date the product is to be released for production.

5. Page. Enter the FMCA page number.

6. Date. Enter the date the page was worked on. Or, enter the revision date, if it is a revised FMCA.

7. Line number. Identify the part for which the FMCA is to be conducted.

8. Cross-reference number. Enter the number if there is a cross-reference with other parts or assemblies.

9. Circuit location. Describe the location of the part on the circuit.

10. Enter the part/component number/name. Enter the appropriate name.

11. Function(s) and Specification(s). Describe the function(s) the part is to perform and the specification(s) required. Make the description as clear and concise as possible. Be sure you include all functions. Include pertinent information about the product specification, such as operating current range, operating voltage range, operating environment, and everything else that is applicable and appropriate.

12. Potential failure mode(s). A failure mode is a design flaw or change in the product that prevents it from functioning property. The typical failure modes are a short circuit, open circuit, leak, loosening. The failure mode is expressed in physical terms of what the customer will experience.

13. System effect. The system effect is what a system or a module might experience as a result of the failure mode. List all conceivable effects, including unsafe conditions or violations of government regulations. A typical system effect is a system shutdown or a failure of a section of the product.

14. Unsafe. Enter 0 for unsafe end product condition.

15. Cause of failure. The ROOT CAUSE — not the symptom — is the real cause. Examples: Insufficient/inaccurate voltage, firmware errors, missing instructions on drawings.

16. Internal or external countermeasures (controls). Identify the controls and/or measures established to prevent or detect the cause of the failure mode. Examples: Perform a derating analysis, perform transient testing, perform specific testing, identify specific inspection and manufacturing specifications.

17. Severity. An estimate of how severe the subsystem and/or the end product will behave as a result of a given failure mode. Severity levels are being scaled from 1 to 10. Number 10 is to be used for a definite unsafe condition, and number 0 is to be used for a negligible severity (nuisance). Usually this rating, at this stage, is very subjective.

18. Base failure rate (λB). A subjective estimate of failure rate (probability of failure in a billion hours). This is also called inherent failure rate.

19. Failure ratio. A subjective likelihood in comparison to the other failure modes. The sum of all failure rates for a part/component should be equal to 10 percent.

20. Effectiveness. A subjective estimate of how effectively the prevention or detection measure eliminates potential failure modes. A typical ranking is the following:

    1   = The prevention or detection measure is foolproof.

    2–3 = Probability of failure occurrence is low.

    4–6 = Probability of occurrence is moderate.

    7–9 = Probability of occurrence is high.

    10   = Very high probability. The prevention/detection measure is ineffective.

21. Risk priority number (RPN). The product of severity, base failure rate, failure mode ratio and effectiveness.

Figure 7-3. A typical design FMCA form.

Operation name (1) _____
Workstation (2) _____
Responsible engineer (3) _____

Subassembly number (4) _____
Supplier (5) _____
Original date (6) _____

Production release date (7) _____
Page (8) _____ of _____
Revised date (9) _____

Line number (10)	Cross-reference number (11)	Circuit location (12)	Enter the part/component number/name (13)	Operational steps (14)	Potential failure mode(s) (15)	Cause(s) of failure (16)	Internal or external counter-measures (controls) (17)	Severity (18)	PPM (19)	Effectiveness (20)	Risk priority number (RPN) (21)

1. Operation name. Enter the name of the operation.
2. Workstation. Enter the name or number of the workstation.
3. Responsible engineer. Enter the name of the responsible engineer.
4. Subassembly number. Enter the subassembly name or number.
5. Supplier. Indicate where the process is performed.
6. Original date. Enter the date that the FMCA is due and/or completed.
7. Production release date. Enter the date the product is to be released for production.
8. Page. Enter the FMCA page number.
9. Revise date. Enter the date of the revision.
10. Line number. Identify the part for which the FMCA is to be conducted.
11. Cross-reference number. Enter the number if there is a cross-reference with other parts or assemblies.
12. Circuit location. Describe the location of the part on the circuit.
13. Enter the part/component number/name. Enter the appropriate name.
14. List all steps of operations in the process. A good tool to use for this is the process flow diagram.
15. Potential failure mode(s). A process-related failure mode is a deviation from specification caused by a change in the variables influencing the process. Examples: Damaged board, misaligned, discolored, missing, bent, etc.
16. Cause of failure. The ROOT CAUSE — not the symptom — is the real cause of the failure. Examples: Transient, human error, machine out of tolerance, ESD equipment failure.

17. Internal or external countermeasures (controls). Identify the controls and/or measures established to prevent or detect the cause of the failure mode. Examples: Verify tooling to its specification, effective incoming inspection, testing, etc.
18. Severity. A subjective estimate of how severe the subsystem and/or the end product will behave as a result of a given failure mode. Severity levels are being scaled from 1 to 10. Number 10 is to be used for a definite unsafe condition, Number 1 is to be used for a negligible severity (nuisance).
19. PPM. Is the percentage failure per 1 million parts.
20. Effectiveness. A subjective estimate of how effectively the prevention or detection measure eliminates potential failure modes. A typical ranking is the following:

   1   = The prevention or detection measure is foolproof.
   2–3   = Probability of failure occurrence is low.
   4–6   = Probability of occurrence is moderate.
   7–9   = Probability of occurrence is high.
   10   = Very high probability. The prevention/detection measure is ineffective.

21. Risk priority number (RPN). The product of severity, PPM, and effectiveness.

Figure 7-4. A typical process FMCA form.

Company: _____
Division/Plant: _____
Department: _____
Process: _____
Operation: _____
Machine: _____
Station: _____

Part Name: _____
Part Number: _____
Control Plan Orig. Date: _____
FMEA Orig. Date: _____
Control Plan Rev. Date: _____
FMEA Rev. Date: _____
Page _____ of _____ Pages

Authorized Control Plan By: _____
Authorized FMEA By: _____
Control Plan Leader: _____
FMEA Leader: _____
FMEA members: _____

Char #	Characteristic Description (product & process)	Spec	TYP / IMP	FAILURE MODE	Effects of Failure	SEV	Causes of failure	OCC	CURRENT CONTROLS	DET	RPN	RECOM. ACTIONS	ACTIONS TAKEN	SEV	OCC	DET	RPN	RESPONSIBLE PERSON	Ctrl. Fact	CLASS	Control Method	TOOL	Gage, desc, master, detail	Other *

* Typical other items may include:

GR & R & Date	C_p/C_{pk} (target) & Date	Reaction plans	Inspection requirements	Sampling requirements

Figure 7-5. Dynamic control plan with failure mode and effect analysis.

108

In the quantitative form, one uses actual statistical and/or reliability data from the processes. This actual data may be historical or current. In both cases, the numerical values are from 1 to 10 and denote set probabilities from low to high, respectively.

The first criterion in the guidelines is occurrence. Under occurrence, we look at the frequency of the failure as a result of a specific cause. The higher the ranking is, the more frequently the failure occurs.

The second criterion in the guidelines is severity. For severity, we assess the seriousness of the effect of the potential failure mode to the customer. Severity applies to the effect and to the effect only. The ranking that is typically given is from 1 to 10, with 1 representing a nuisance and 10 representing a major noncompliance to government regulations or safety item.

The third criterion in the guidelines is detection. For detection, we assess the probability of a failure reaching the customer. The higher the ranking is, the more likely your current controls or design verification did not detect the failure or contain the failure within your organization. Therefore, the likelihood of your customer receiving a failure is increased.

A general complaint about these guidelines is frequently heard in regards to consensus; for example, How can a group of people agree on everything? The answer is that they cannot, but that does not preclude the use of consensus to decide. The following is one of the ways to use it.

If the decision falls in an adjacent category, the decision should be averaged. If, on the other hand, it falls in more than an adjacent category, you should stick to the consensus process. Problems at this point may be that someone on the team may not understand the problem, some of the assumptions may have been overlooked, or maybe the focus of the team has drifted.

If no consensus can be reached by a reasonable discussion among the team members, then traditional organizational development (team dynamic and problem-solving techniques) must be used to resolve the conflict. Under no circumstances should a mere agreement or majority be pushed through so that an early completion may take place. Remember that all FMEAs are time consuming, and the participants, as well as their management, must be patient in order to achieve the proper results.

Risk priority number

The risk priority number (RPN) is the product of occurrence, severity, and detection. The value should be used to rank the concerns of the design and/or process. By itself, the RPN has no other value or meaning.

A threshold for pursuing failures is an RPN equal to or greater than 50 based on a 95% confidence (although there is nothing magical about 50). We can identify this threshold by any statistical confidence. For example, if 99% of all failures must be addressed for a very critical design, what is the threshold? There are 1000 points available ($10 \times 10 \times 10$) from occurrence, severity, and detection; 99% of 1000 is 990. So, anything equal to or over 10 as an RPN must be addressed. On the other hand, if we want a confidence of only 90%, then 90% of 1000 is 900. So, anything equal or greater than 100 as an RPN must be addressed. Please note that this threshold is organization dependent and can change with not only the organization, but the product and/or the customer.

If there are more than two failures with the same RPN, then we use severity, then detection, and finally occurrence as the order of priority. Severity is used first because it deals with the effects of the failure. Detection is used over the occurrence because it is customer dependent, which is more important than just the frequency of the failure. (The reader must also understand that nothing prevents action on every failure identified. However, if that is the desired course, then the FMEA is superfluous. After all, we use the FMEA to identify and set priorities for the importance of failure and act accordingly.)

Recommended action

When the failure modes have been rank ordered by RPN, corrective action should be first directed at the highest ranked concerns and critical items. The intent of any recommended action is to reduce the occurrence, detection, and/or severity rankings. Severity will change only with changes in design; otherwise, more often than not, the reductions are expressed in either occurrence and/or detection. If no actions are recommended for a specific cause, then this should be indicated. On the other hand, if causes are not mutually exclusive, a DOE recommendation is in order.

In all cases for which the effect of an identified potential failure mode could be a hazard to manufacturing or assembly personnel, corrective actions should be taken to prevent the failure mode by eliminating or controlling the cause(s). Otherwise, appropriate operator protection should be specified.

The need for ranking specific, positive corrective actions with quantifiable benefits, recommending actions to other activities, and following up all recommendations cannot be overemphasized. A thoroughly thought-out and well-developed FMEA will be of limited value without positive and effective corrective actions. It is the responsibility of those dealing with all affected activities to implement effective follow-up programs to address all recommendations (Chrysler, Ford, and General Motors, 1995b).

The "process" of doing an FMEA

To do an FMEA effectively, one must follow a systematic approach. The recommended approach is an eight-step method that facilitates the system, the design, the process, and the service FMEAs.

1. *Brainstorm.* Try to identify which direction you want to go: system, design, process, or service. What kind of problems are you having with a particular situation? Is the customer involved, or are you pursuing continual improvement on your own? If the customer has identified specific failures, then your job is much easier because you already have the direction identified. On the other hand, if you are on your own, the brainstorm and/or the cause-and-effect diagram may prove to be the best tools to identify your direction of attack.

2a. *Block diagram.* (used only for system and design). Make sure everyone on the team is on the same wavelength. Does everyone understand the same problem? The block diagram will focus the discussion. If nothing else, it will create a baseline of understanding for the problem at hand.

2b. *Process flow chart.* (used only for process and service). Make sure everyone in the team is on the same wavelength. Does everyone understand the same problem? The process flow chart will focus the discussion. If nothing else, it will create a baseline of understanding for the problem at hand.

3. *Set priorities.* Once you understand the problem and its relationship to other problems, the team must begin the plan for action. For example, for one problem there may be subproblems or areas that need to be addressed first. A thorough study of the block diagram or the process flow chart will indicate—more often than not—the appropriate action. This is the stage at which we really begin to focus on the problem.

4. *Data collection.* Begin to collect data on the failures and start filling out the appropriate form.

5. *Analysis.*   Focus on the data and perform the appropriate analysis. Everything is fair, provided it is appropriate and applicable. Here, QFD (Quality Function Deployment) DOE, another FMEA, SPC, and anything else may be used to provide the appropriate results.

6. *Results.*   Based on the analysis, the results are derived. The results must be data driven; nothing else will do.

7. *Confirm/evaluate/measure.*   Once the results have been recorded, it is time to confirm, evaluate, and measure the success or failure of the recommended action taken. This evolution takes the form of three basic questions:

- Are we better off than before?
- Are we worse off than before?
- Are we the same as before?

8. *Do it all over again.*   Regardless of how you answer Step 6 and Step 7, you must pursue improvement all over again because of the continual improvement philosophy. Your long-range goal is to eliminate completely all failures; your short-term goal is to minimize your failures, if not eliminate them. Of course, the perseverance for those goals has to be taken into consideration in relationship to the needs of the organization, cost, customer, and competition.

For detailed coverage of FMEA, see Stamatis (1995a) and Chrysler, Ford, and General Motors (1995b).

Fault tree analysis (FTA)

FTA may be used to identify potential causes of failure that a product, system, or process may experience. This analysis differs from FMEA in that FTA starts at the top level of a system and works down to possible causes. The general steps are:

- Define the top level fault event.
- Start the analysis at the highest level in the system.
- Analyze down to independent causes.
- Utilize logic symbols.

The FTA graphically and logically depicts all the events or combinations of events leading to the top-level fault event. It shows what was considered in the analysis and how the analysis logic proceeded. FTA can also provide a format for a troubleshooting guide for diagnostics when problems are encountered during product use.

Design reviews

Design reviews are an integral part of the design and development process. It is a formalized, documented, and systematic management process through which both the machinery supplier and the user review all technical aspects. The review objectives depend on the stage of the design phase. Typical objectives are shown in Table 7-2.

Phase 3. Build and install

Tolerance studies

Routinely, suppliers must conduct studies to ensure that stacking in electrical and/or mechanical systems does not cause equipment failure or premature wear.

Table 7-2. **Typical Design Review Objectives**

Design phase	Review objectives
Concept	To review the feasibility results and the design approach
Development and design	Validate the capability of the design for all technical requirements
	Verify that the design and all pertinent analyses are complete and accurate
Build and install	Resolve issues that resulted from testing in machine build and runoff
	Conduct failure investigation of problem areas for continuous improvement

Adapted from Society of Automotive Engineers and National Center for Manufacturing Sciences, Incorporated, Reliability and Maintainability Guideline for Manufacturing Machinery and Equipment, National Center for Manufacturing Sciences, 1993. Ann Arbor, MI. Used with permission.

Stress analysis

Stress analysis uses numerical analysis to determine the relationship between the strength of the component and the stress induced by the environment under worst-case conditions.

Dedicated reliability testing

Dedicated reliability testing is done to verify attainment of the R&M requirements. The selection and the duration of the testing should be agreed on by both user and supplier.

Root cause/failure analysis

Machinery suppliers are responsible for ensuring that root cause analyses of equipment failures are performed; the results will be fed back to the user so that the user and supplier can jointly determine how to best resolve any deficiencies. The process should be documented.

Phase 4. Operation and support

Failure reporting

Suppliers are responsible for reporting all failures and conducting a failure mode analysis. The focus of the analysis should be to determine the root cause.

Analysis and corrective system

The machinery supplier should develop a means for summarizing and analyzing failure data obtained at both supplier and user plants as a means of promoting R&M improvement activities.

Phase 5. Documentation/conversion

The specific documentation/conversion records required depend on the tool and/or equipment built.

SYSTEM EFFECTIVENESS

The essence of R&M is to establish system effectiveness. System effectiveness is often expressed as one or more figures of merit representing the extent to which the system is able to perform the intended function. The figures of merit used may vary considerably depending on the type of system and its mission requirements; it should consider the following:

1. System performance parameters as defined by the customer and agreed by the supplier.
2. Availability is the measure of the degree to which a system is in the operable and committable state at the start of a mission when called for at an unknown random point in time. This is often called "operational readiness." Availability is a function of operating time (reliability) and downtime (maintainability/supportability).
3. Dependability is the measure of the system operating conditions at one or more points during the mission, given the system condition at the start of the mission (i.e., availability). Dependability is a function of operating time (reliability) and downtime (maintainability/supportability).

A combination of the three measures represents the system effectiveness aspect of total cost effectiveness (Blanchard, 1986).

LIFE COST CYCLE

Life cost cycle (LCC) involves all costs associated with the system life cycle, including the following:

1. *R&D cost.* The cost of feasibility studies, system analyses, detailed design development, fabrication, assembly and test engineering models, initial system test, and evaluation and associated documentation.

2. *Production and construction cost.* The cost of fabrication, assembly, and test of operational systems (production models); operation and maintenance of the production capability and associated support requirements (e.g., test and support equipment development, spare/repair parts provisioning, technical data development, training, entry of items into the inventory, facility construction, and so on).

3. *Operation and maintenance cost.* The cost of sustaining operation, personnel and maintenance support, spare/repair parts and related inventories, test and support equipment maintenance, transportation and handling, facilities, modifications and technical data changes, and so on.

4. *System retirement and phaseout cost.* The cost of phasing the system out of the inventory due to obsolescence or wearout and subsequent equipment item recycling and reclamation as appropriate.

Life cycle costs may be categorized in many different ways, depending on the type of system and the sensitivities desired in cost-effectiveness measurement (Blanchard, 1986).

8

Miscellaneous topics concerning the tooling and equipment industry

As in every industry, some issues and concerns present themselves uniquely in the tooling and equipment industry. This chapter covers some of these special topics from an overview perspective. The intent is to sensitize the reader to the issue rather than give an exhaustive explanation of it. The topics covered are Robust Development and Design (RDD), Geometric Dimensioning and Tolerancing (GD&T), short-run Statistical Process Control (short-run SPC), measurement system analysis (MSA), capability analysis, error/mistake-proofing, and continuous improvement.

ROBUST DEVELOPMENT AND DESIGN

Robust Development and Design is one of the most misunderstood concepts in the tooling and equipment industry. To understand the concept, one must understand its terminology. First, the term *robust* means a product/process that provides consistent, high-level performance despite a wide range of changes in both customer expectations and production conditions. In essence, then, a robust design is a process for optimization.

Second, *development* provides the tools and skills needed to begin developing robust, generic technology that can be applied efficiently across a broad range, or family, of present and future products. In essence, development is a corporate strategy that enables organizations to reduce existing product development cycles. It does this by bridging the gap between research and development (R&D), invention and innovation, and product/process development. Finally, it creates a system for rapid deployment and commercialization of new products.

Third, *design* is the phase of product development that makes the design for specific items robust and insensitive to uncontrollable sources of variation and user conditions.

One way that RDD is applied in the tooling and equipment industry is through statistical thinking for problem solving. Statistical thinking is the approach of determining and using appropriate statistics to make the best existing process better. This thinking must not be confused with that of SPC. SPC aims to achieve the best output from an existing process in a consistent, repeatable, and predictable manner. Statistical thinking, on the other hand, carries quality improvement a step further by aiming to make the best even better. Statistical thinking is the strategic application of statistical methods to reduce process output variation beyond what can be accomplished with traditional SPC. The process of using this statistical thinking may be summarized in five steps:

1. Define the problem.
2. List the variables.
3. Select priorities for variables.
4. Evaluate variables from previous step and their interaction, if applicable.
5. Optimize variables by using DOE or even a simulation package within the CAD (Computer-Aided Design) system of your company.

Another approach is via DOE—classical designs. DOE is a way to structure testing programs efficiently to obtain information. It can aid in the identification of single factors or a combination of factors (interaction) with an effect on the desired output. It can be used to identify the factors that produce the most desirable/stable output in the presence of uncontrollable conditions.

A third approach is DOE—the Taguchi approach. Taguchi approaches quality from a different point of view. For him, quality is a holistic view relating cost to "loss" in dollars to society as a whole. The application of Taguchi's methodology to the tooling and equipment industry is of major interest primarily because it deals with both parameter design and tolerancing design. For more information on robust designs, see Taguchi (1987). For a good understanding of statistical thinking, see Britz, Emerling, and Hare (1996).

GEOMETRIC DIMENSIONING AND TOLERANCING

GD&T is a technical database through which product design and production personnel can communicate via drawings to provide a uniform interpretation of the requirements for making a product. It replaces confusing and inconsistent notes and datum lines with standard symbols that refer to a universal code, the American Society of Mechanical Engineers (ASME) Y14.5M-1994 Dimensioning and Tolerancing Standard (ASME, 1994), an international symbolic engineering language. This standard improves cross-functional communication and helps avoid costly changes to drawings after initial release.

GD&T allows design engineering to communicate the requirements of a part more clearly and should eliminate the need for implied and/or ambiguous part definition. It is beyond the scope of this book to delineate all the characteristics of GD&T. However, because of the important role it plays in translating the customer requirements through the drawing(s) in the tooling and equipment industry, the following short list of characteristics or concepts is presented.

Bonus tolerance. When the maximum material condition (MMC) or the least material condition (LMC) modifiers are used with a geometric specification as a produced part's size departs from its stated size limit, an additional amount of the geometric tolerance is permitted.

Composite position. Using this design tool, it is possible to allow greater tolerance on a pattern of holes to the boundary of a part while maintaining tighter control of the spacing within the pattern of holes.

Datum reference frame. The datum reference frame defines the location and/or orientation of a part feature in relation to other specific part features. The datum reference frame is made up of three mutually perpendicular planes that correspond to the X, Y, and Z axes. The concept replaces the use of implied data, which often leads to erroneous interpretations. Data may then be used by production, but is intended to identify correct inspection set-up procedures.

Diametral tolerance zone. A concept that creates a cylindrical tolerance zone for the axis of round holes, pins, studs, and shafts. While there are several applications for this type of tolerance, it is expressly intended to allow additional tolerance for round mating features.

Functional specification/gaging. If the MMC modifier is used, we are able to make a functional design specification. That is, bonus tolerance is permissible as long as the part will perform its intended function. We are also able to verify part conformance by a simple go/no-go gaging process when feasible.

GD&T Controls. There are 13 control symbols that are used to define various characteristics and replace the typical title block or note specifications found in drawings. All of these 13 are clearly defined in the standard, thereby eliminating any ambiguity.

Material conditions. There are drawing symbols that are used to identify three material conditions (LMC; MMC; regardless of feature size, RFS.) These material conditions, when appropriately applied, more clearly define the function of a part and can allow additional tolerance or datum shift when permissible in the design.

Projected Tolerance Zone. A concept that is generally applied to tapped holes and dowel pinholes to ensure a noninterference fit during assembly. Without this design consideration, it is difficult to obtain tight-fitting or controlled assembly conditions due to orientation limitations.

Rules. There are three rules; two deal with the application of the material condition modifiers. Rule 1 is by far the most significant and delineates the concept of form also being controlled by the size limits.

Virtual condition. A term used to describe the worst-case accumulative effect of size and any applicable geometric tolerances. This condition is used in mating part design to ensure part fit, to do stack-up analysis, and also to design functional gaging and fixturing.

Zero tolerance. Using zero tolerance with the MMC modifier in a geometric specification assures you that any part you buy will function properly. For more information on GD&T, see ASME Y14.5M-1994 (ASME, 1994).

SHORT-RUN STATISTICAL PROCESS CONTROL

In the 1980s, SPC was reintroduced in the manufacturing industry with both enthusiasm and very good results in process improvement. However, when SPC was tried in the tooling and equipment industry, it met with difficulty and provided some questionable results. These results were experienced because (1) SPC was misunderstood, and (2) in tooling and equipment organizations there is an issue of the "short run."

First, let us examine the issue of misunderstanding. When someone hears the term *SPC*, charts and statistics are automatically assumed and envisioned. That may or may not be correct. SPC is a bundle of tools to be used in identifying, defining, correcting, evaluating, and monitoring a process for some kind of improvement. The improvement is organization dependent and may be defined differently for each organization, as well as each process. Typical improvement expectations are reduced waste, improved efficiency, improved cycle time, and so on.

To measure the improvement in an organization, one may use qualitative and/or quantitative tools (both are part of SPC). Some of the qualitative tools are brainstorming, affinity charts, process flow chart, and cause-and-effect diagrams. Some of the quantitative tools are check sheets, control charts, scatter plots, general statistics, and advanced statistics.

What is important about both categories of tools is the fact that they can be used in any organization given the appropriate and applicable data. Therefore, in the tooling industry there are situations for which SPC is appropriate and quite useful. By the same token, however, there are some limitations, which brings us to the second point—limited production.

A short run is an environment in which there are a large number of jobs per operator in a production cycle—typically a week or month—with each job involving different products. In essence, a short run involves the production of very few products of the same type, termed a *limited production run*. An extreme case of a short production run is the one-of-a-kind product.

When a short run is used, traditional SPC is not applicable. Therefore, the analysis should focus on the process itself as opposed to the product. When that happens, modifications are in order. For example, since short runs involve less than the recommended number of pieces for capability, the acceptability criteria are usually modified.

It is beyond the scope of this book to identify and discuss in detail all of the specific tools used in short-run SPC. The reader is encouraged to see Griffith (1996), Pyzdek (1992), Hillier (1969), and others. Some of the common tools are:

traditional X-bar and R charts

traditional individual (X) chart

traditional individual (X) chart and moving range

traditional attribute charts (p, c, np, and u)

code value charts (evaluating the deviations of the target, rather than the actual value)

stabilized control charts (multiple characteristic charts with independent units)

the exact method as defined and developed by Hillier (1969)

standardized control charts

demerit control charts

It is important that these tools be applied when the appropriate data have been defined and collected.

MEASUREMENT SYSTEM ANALYSIS (MSA)

Measurement plays a significant role in helping a facility accomplish its mission. Therefore, the quality of the measurement systems that produce those measurements is important. For example, too much variation in a measurement system being used for SPC may mask important variations in the production process. A typical measuring system in any tooling and equipment organization should include

design and certification

capability assessment (over time)

mastering

gage repeatability and reproducibility (gage R&R)

operational definition

control

repair and recertification

Perhaps the most important of these is gage R&R. It is a technique to measure inherent variation in measuring equipment and variation between gage operators. It does reveal how much of the specification limit spread is used by measurement nonrepeatability. The gage R&R establishes gage accuracy, repeatability, reproducibility, stability, and linearity. Gage accuracy is the difference between the observed average of measurements and the master value. Repeatability is the measure of the gage to repeat a given measurement. Gage repeatability is the variation in measurement obtained with one gage when used several times by one operator while measuring the identical characteristics on the same parts. Reproducibility is the measure of the ability of people to repeat a given requirement. Gage reproducibility is the variation in the average of the measurements made by different operators using the same gage when measuring identical characteristics on the same parts. Stability is the total variation in the measurements obtained with a gage on the same master parts when measuring a single characteristic over an extended time period. Sometimes stability is also referred to as *drift*. Linearity is the difference in the accuracy values through the expected operating range of the gage.

It is very important to remember that during gage calibration, linearity and accuracy should be performed before assessing gage R&R.

$$RR = (AV^2 + EV^2)^{1/2}$$

Gage R&R quantifies measurement system variation (RR) attributable to equipment variation (EV) and appraiser variation (AV). Repeatability is equipment variation, and reducibility is appraiser variation. Measurement system variation may be expressed as a percentage of tolerance, such as

$$\%Error = 100\% \ RR/\text{Tolerance}$$

Finally, in evaluating the gage R&R, the following are guidelines for acceptable %Error: under 10%, gage is acceptable; 10%–30%, gage may be acceptable depending on the importance of the application, cost of gage, cost of repairs, and so on; over 30%, gage needs improvement and make every effort to identify and correct causes of variation or use a different measurement system. For more information, see Chrysler, Ford, and General Motors (1995c).

CAPABILITY ANALYSIS

From personal experience over the last 30 years in the field of quality, I have observed that a large portion of the cost of poor quality is caused by manufacturing processes that are incapable of producing parts to the specifications. This brings us to the tooling and equipment industry since it provides the machinery, tooling, and fixturing. What is interesting, however, is that quite a few of the tooling and equipment suppliers will justify their delivery of incapable machines and/or equipment by claiming that the specifications were misunderstood, the material variation was greater than predicted, or the machine was the first of its kind, and so on.

To avoid such a predicament, a tooling, equipment, and fixturing supplier must understand the concept of capability and what the requirements are for documenting such capability. Again, it is beyond the scope of this book to address the complete topic of capability; a short review of the topic follows.

A capability analysis evaluates the ability of the process to meet specifications. Stability, of course, is generally a prerequisite for calculating capability. There are several indices that can be used to quantify process capability. All are calculated based on data collected from the process. Selection of a capability index depends in part on the type of data collected. Examples are the following:

Variable data	Attribute data
C_p	First-time capability (FTC; percentage good)
C_{pk}	Average percentage defective
	Average defects per sample
	Average defects per unit

C_p is sometimes referred to as the process potential because it considers only process variation (spread), not process location. It predicts the potential performance if the process was perfectly centered. C_{pk} accounts for both variation and location. The higher the C_p or C_{pk} number is, the better the process capability will be. For a perfectly centered process, the C_{pk} and C_p are equal.

In the tooling and equipment industry, because of limited samples, the designations P_p and P_{pk} are sometimes used instead of C_p and C_{pk} when the process spread is calculated using total sample deviation rather than the standard deviation estimated from a control chart (R-bar/$d2$). This alternative calculation is also particularly useful prior to the start of full production when there are insufficient data to generate meaningful control charts. P_p and P_{pk} are referred to as process performance.

The FTC is the ratio of correctly processed items to the total number of items processed, usually expressed as a percentage. Both the FTC and the

average percentage defective have the ability to be rolled up mathemati-
cally so that the performance of individual operations or zones can be
combined to calculate total plant FTC or average percentage defective. C_p
and C_{pk} values, on the other hand, can be calculated only for a single
process characteristic. It is not valid, for instance, to calculate a C_{pk} if the
process is not stable, and it certainly is not valid to combine C_{pk}'s into an
average C_{pk}.

Another way to calculate the capability is through a graphical represen-
tation using a normal probability paper. Usually, the graphic provides a
way to present a visual picture of the spread found among a sample group
of parts. A straight line or close to it is interpreted as the process being
capable. If the line is not straight, then there is unusual variation in the
process, and it must be removed through problem-solving techniques.

Once the basic understanding of capability has been communicated in
the organization, then it is imperative that the documentation of such an
activity be demonstrated. This demonstration must be done at least
through the following:

- Critical characteristics and required capability ratio
- Specified source for material
- Quantity and cycles to be run
- Person who will make the evaluation
- Source and type of gaging
- Measurement methods
- Process requirements

After the equipment is built, the supplier performs the capability study.
Major items the supplier should report to the customer are contained in a
"capability document," which should include a summary of the measurements and explanations of the results. The items addressed for explanations might include the following:

evidence of drift

shape of the process distribution

any adjustments, stoppages, or interruptions that occurred during
the run

for multiple-stream processes, identify fixture, hand, pallet, or cavity

gage variability and calibration

existence of statistical control for mean and variability process

if the characteristics satisfy the capability requirements

if the process is not capable, recommendations for type of corrective action needed to fix the machine.

Minimum recommended criteria for machine and process capability acceptance are a machine (short-run) capability C_{pk} of 1.67 and a process (long-run) capability P_{pk} of 1.25. For more information, see Chrysler, Ford, and General Motors (1995e), IBM (1986), and Ford (1985).

ERROR/MISTAKE-PROOFING

Error-proofing is a means to design potential failure modes out of a product or process. In essence, you design something with the intent that the mistake cannot be made. By identifying potential problems through a design or process Failure Mode and Effect Analysis (FMEA), corrections can be made to the design to eliminate these potential problems.

Mistake-proofing is any change to the operation that helps the operator reduce or eliminate mistakes and/or provides immediate feedback and corrective action should mistakes occur. The purpose is to prevent mistakes from continuing in the process or being received by the customer.

For more information, see Shingo (1986).

CONTINUOUS IMPROVEMENT

The essence of continuous improvement is a strategic, integrated management system for achieving customer satisfaction that uses measurable practices to continuously reduce waste within an organization. These practices have to do with management, processes, and people.

Usually, management ensures continuous improvement as a living corporate strategy by providing support, direction, and the allocation of resources through programs, systems, and processes. Middle management deploys and implements goals, as directed by senior management, by providing intensive education and training, as appropriate, for the utilization of employee skills and tools for problem solving and problem prevention. The fundamental requirement in continuous improvement and the commitment of management is the degree of empowerment by all employees.

Processes, on the other hand, are a series of steps to accomplish an objective. Furthermore, all processes may have the following components: personnel, machines, material, method, measurement, and environment.

Whatever components a process has or what its objective is, traditionally all processes create waste. By understanding waste, employees determine ways to reduce or eliminate waste in any of the six components or in any combination of these components. The essence, then, of continuous improvement in a process is to reduce variation in that process with the goal of reducing the waste.

Finally, people, the ones in the trenches with process measurables, are responsible for identifying, reducing, or eliminating wasteful activities. Participation by the full workforce is reinforced as individual workers and teams excel in an environment in which people work together to increase personal satisfaction. The essence of continuous improvement from the people's perspective is to take ownership and manage their functions in such a way that everyone learns to become efficient.

Appendix A: Quality manual

In this appendix, we provide a sample quality manual. The manual is based on the TE Supplement requirements. However, the reader is warned that this manual is only an example and should be used as a guideline in developing your own quality manual.

Document No: QM XYZ 001
Revision: 00
Date: March 1, 1997
Supersedes: New
Section: 0
Page 1 of 1
Authorized By: _____

XYZ, Inc.
QUALITY MANUAL
Based on the TE Supplement
(Sample Manual)

Document No: QM XYZ 001

Title: XYZ's Quality System Practices and Controls: Revision: 00
        Table of Contents   Date: March 1, 1997
                                                                    Supersedes: New
                                                                        Section: 0
                                                                       Page 1 of 2

Title: XYZ's Quality System Practices and Controls:
    Table of Contents

Document No: QM XYZ 001
Revision: 00
Date: March 1, 1997
Supersedes: New
Section: 0
Page 2 of 2

Section II
Sector-Specific Requirement
II.1 Purpose
II.2 Continuous Improvement
II.3 Manufacturing Capabilities

Section III
Customer-Specific Requirements
III.1 Chrysler Requirements
III.2 Ford Requirements
III.3 General Motors Requirements

5.0 Appendices
(May include the following even though these items are not included here)
5.1 XYZ Organizational Chart
5.2 Cross-Reference Between QS-9000 Requirements and the XYZ Quality System
5.3 Cross-Reference Between Individual Automotive Requirements and the XYZ
    Quality Manual
5.4 Miscellaneous
5.5 Cross-References among Quality Manual, Procedures and Work Instructions

	Document No: QM XYZ 001
Title: XYZ's Quality System Practices and Controls:	Revision: 00
Introduction	Date: March 1, 1997
	Supersedes: New
	Section: 1.0
	Page 1 of 1

1.0 Introduction

1.1 XYZ Quality Policy

XYZ will provide products and services that meet or exceed the expectations of our customers (internal and external) through our total commitment to continuous improvement in everything we do. We are committed to world-class quality for everything we do, from design to assembling to shipping to customer follow-up. To implement this policy, we are educating all our employees on how their performance contributes to product quality, and we encourage measurement characteristics to track our improvement.

In addition, we are committed to empowering all our employees with the appropriate authority and responsibility to "get" the job done the first time within their own jurisdiction.

1.2 Scope of the XYZ Quality Manual

The XYZ Quality Manual describes the policies and quality system structure employed by XYZ to ensure that the design, assembly, and delivery are in good order. The quality system was created to reflect a quality program consistent with the ISO 9001 standard and the automotive requirements known as QS-9000 and the TE Supplement.

1.3 Company Profile

XYZ was founded in 1947 with the sole purpose of supplying the automotive industry with tool and die supplies. The company started out with 10 employees and by 1994 had grown to 1168. The gross sales went from a low of 10MM to 968MM in 1996.

The company has grown not only in financial terms and employees, but also in product line, as well as facilities. XYZ operates 15 plants throughout the United States, Canada, Mexico, Germany, and Brazil and has extended the product line to include automotive fixtures.

	Document No: QM XYZ 001
Title: XYZ's Quality System Practices and Controls: References	Revision: 00
	Date: March 1, 1997
	Supersedes: New
	Section: 2.0
	Page 1 of 1

2.0 References

2.1 Quality Standards

The following quality standards and/or guidelines have been used in developing this quality manual:

- ISO 9001 Quality Systems—Model for Quality Assurance in Design/ Development, Production, Installation, and Servicing

- ISO 9004 Quality Management and Quality System Elements Guidelines

- Quality System Requirements: QS-9000

- TE Supplement to QS-9000

- ISO/DIS 10013 guidelines for developing quality manuals

2.2 XYZ Quality System Documents

This manual contains only the policies of XYZ, Incorporated for the design and assembly of high-quality products. In case more detail is needed, the reader is encouraged to consult the procedures and instructions manuals in the XYZ Quality Documentation System. The appendices of this manual contain reference tables and requirements that correspond with and demonstrate compliance to these standards. Because documentation is constantly changed to reflect the requirements of the customer, the reader may consult the XYZ quality assurance (QA) documentation department for an updated listing of documents.

Document No: QM XYZ 001
Title: XYZ's Quality System Practices and Controls: Revision: 00
Quality System Terminology   Date: March 1, 1997
Supersedes: New
Section: 3.0
Page 1 of 1

3.0 XYZ Quality System Terminology

3.1 Terms and Definitions

No special terminology or jargon is used. If there is a need for explanation of
certain words, the official interpretation is based on the ANSI/ASQC A8402
standard and the R&M guidelines.

3.2 Standard Abbreviations

Within XYZ, we use many abbreviations, and the reader is encouraged to consult
our booklet, which is found in every office of the company, on acronyms and
abbreviations.

Document No: QM XYZ 001
Revision: 00
Date: March 1, 1997
Supersedes: New
Section: 4.1
Page 1 of 3

Title: XYZ's Quality System Practices and Controls:
Management Responsibility

4.0 Quality System Practices and Controls

4.1 Management Responsibility

4.1.1 Quality Policy Requirements

It is the responsibility of XYZ management to provide the personnel and resources necessary to maintain the requirements of the XYZ Quality Policy and the XYZ Quality Manual.

Specifically, management will

• Provide quality products and services by striving to exceed the defined needs and expectations of our customers.

• Develop a quality system based on ISO 9001 and the QS-9000 standard to foster continuous process improvement and problem prevention instead of problem detection. As is appropriate and applicable, the TE Supplement will be followed.

• Define and implement our quality system based on employee empowerment and a commitment to excellence.

• Give all employees the training and support needed to provide quality products and services to all customers.

• Communicate our mission and quality objectives to all employees and assign individual responsibilities for quality and accountability.

• Use statistical methods when appropriate to monitor quality performance and isolate major problems for immediate solution.

• Establish and maintain a working environment that supports the production and delivery of high-quality products and services.

• Form relationships with both customers and suppliers that will improve quality in all aspects of product usage and purchased parts.

• Establish objectives that include reliability, maintainability, and durability.

Document No: QM XYZ 001
Revision: 00
Date: March 1, 1997
Supersedes: New
Section: 4.1
Page 2 of 3

Title: XYZ's Quality System Practices and Controls:
   Management Responsibility

4.1.2 Organization

4.1.2.1 Responsibility and Authority Requirements

XYZ has an organizational structure with established levels of responsibility and authority that will provide quality products to our customers and reduce conflicts of interest among departments and functions. A detailed organizational chart is provided in the appendix.

The organizational structure allows independence and authority of the quality function, an essential element for ensuring that quality products are delivered to our customers.

Another important element of the organizational chart is the fact that all the employees are responsible for their own quality and have the freedom, responsibility, and duty to identify and document quality problems for corrective actions that prevent the occurrence of nonconformances in the product/service, process, or quality system.

4.1.2.2 Resources

All resource requirements are defined in the procedures and instructions manual as needed. As for their ability to perform their designated tasks, their qualifications may be found either in the training department and/or their own respective departments.

4.1.2.3 Management Representative

The management representative has the authority and responsibility to ensure that the quality system is effectively implemented and maintained in accordance with the ISO/QS requirements and for reporting to the company's chief executive on the performance of the quality system at the time of the management review.

4.1.2.4 Organizational Interfaces

XYZ is committed to the philosophy of teams and, as such, is always encouraging cross-functional and multidisciplinary teams for continuous improvement.

4.1.3 Management Review

The quality system at XYZ is designed to satisfy the requirements of both ISO 9001 and QS-9000. Management, therefore, will review the system at least once every six months to ensure conformance and effectiveness. The process and the results of the reviews will be documented.

4.1.4 Business Plan

The senior vice president is responsible for ensuring that business goals and plans are implemented and adhered to as described in XYZ's Business Plan. The Business Plan details short- and long-term goals for specific areas of the business, including

Market-related issues
Financial planning and cost

Document No: QM XYZ 001

Title: XYZ's Quality System Practices and Controls: Revision: 00
Management Responsibility   Date: March 1, 1997
                                                          Supersedes: New
                                                             Section: 0
                                                            Page 3 of 3

Growth projections
Plant/facilities plans
Cost objectives
Human resource development
R&D plans, projections, and projects with appropriate funding
Projected sales figures
Quality objectives
Customer satisfaction plans
Key internal quality and operational performance measurables
Health, safety, and environmental issues

The senior vice president is also responsible for ensuring that the Business Plan is developed, documented, tracked, revised, reviewed, and communicated as defined in the appropriate procedure. Goals and plans are developed as a result of the analysis of company-level data, competitor and other benchmark data, customer satisfaction data, and current and future customer needs and expectations.

4.1.5 Analysis and Use of Company-Level Data

The quality staff are responsible for ensuring that trends in quality, current levels of quality for key product and service features, and key indicators for operational performance are documented in the quarterly Business Performance Report. The information documented in these reports is used as an input to the tracking, analysis, and subsequent revision of the business plan.

To make sure that data are used, each department manager prepares monthly "performance" reports that document key internal quality and operational indicators. The reports are submitted to senior management for review, and become key inputs to the Business Performance Report. The manager of quality assurance, in consultation with the department head, initiates a Corrective Action Procedure and/or Preventive Action Procedure, as appropriate, based on the results of the monthly reports in order to ensure that prompt action is taken on customer-related problems and that weaknesses in the quality system are addressed.

4.1.6 Customer Satisfaction

The marketing department is responsible for documenting and tracking trends and current levels of customer satisfaction according to the Determination, Reporting, and Review of Customer Satisfaction Procedures. This responsibility includes determining customer satisfaction for both immediate and end users and collecting competitive and other benchmark data for comparative purposes. The results of these activities are key inputs to the Business Performance Report and Business Plan.

Related Documentation: (Place here all related documentation if appropriate and applicable.)

Title: XYZ's Quality System Practices and Controls:
         Quality System

Document No: QM XYZ 001
Revision: 00
Date: March 1, 1997
Supersedes: New
Section: 4.2
Page 1 of 1

4.2 Quality System

4.2.1 General

XYZ's quality system is defined through four-tier documentation. Its purpose is to ensure that all products and services conform to the requirements as defined by our customers and our own management. The four levels of our system are

- Quality Manual—Defines the policy and companywide structure and methods for maintaining the quality system.

- Procedures—Define who does what, when it is done, and what documentation is used to verify that the quality event was performed as required.

- Work instructions—Define the detail of the procedures.

- Records and forms—Records provide the assurance that the required product or service was achieved. Forms, on the other hand, refers to tags, labels, stamps, printed sheets, and all other means to identify the status of specific material, equipment, and other items in the organization.

4.2.2 Quality System Procedures

XYZ maintains documented quality system procedures to satisfy the ISO, QS, and the TE requirements and to meet the company's need to manage and control the quality system effectively.

4.2.3 Quality Planning

The quality assurance department is responsible for performing quality planning to define the specific quality practices throughout the XYZ company.

The manager of quality assurance has the direct responsibility and authority to define, implement, and maintain the quality system. That responsibility extends, but is not limited, to control plans as defined in APQP and Control Plan, special characteristics, the use of cross-functional teams, feasibility reviews, and the use of FMEAs.

XYZ defines special characteristics as those characteristics that may affect safety or compliance with government regulations, are closely tied to customer satisfaction, or require special monitoring and control for any other reason. It is XYZ's responsibility to familiarize their suppliers with the appropriate identification and use of the special characteristics.

XYZ utilizes and understands the implication(s) of using APQP and R&M through the life cycle process.

Related Documentation: (Place here all related documentation if appropriate and applicable.)

Title: XYZ's Quality Systems Practices and Controls: Revision: 00
      Contract Review   Date: March 1, 1997
Supersedes: New
Section: 4.3
Page 1 of 1

4.3 Contract Review

4.3.1 General

Contract reviews are conducted, documented, and reviewed as appropriate through-out the XYZ company.

4.3.2 Review

The marketing and purchasing vice presidents have the primary responsibility for ensuring that all contracts are reviewed prior to submission and to ensure that they are adequately defined and documented. Furthermore, at this stage of the review, the engineering and/or quality department may be consulted as to the capability of XYZ to fulfill the order.

In the case of verbal orders, XYZ's personnel ensure that the order requirements are agreed before their acceptance. Appropriate procedures exist for such cases.

4.3.3 Amendment to a Contract

It is the marketing department's responsibility to coordinate all amendments to the contract with the customer and XYZ's personnel. When changes do occur, they are reviewed, approved, and communicated to the appropriate personnel within the XYZ company.

4.3.4 Records

All information pertinent to contract reviews are maintained and documented by company name, order number, and customer name and/or code.

Related Documentation: (Place here all related documentation if appropriate and applicable.)

Document No: QM XYZ 001

Title: XYZ's Quality Systems Practices and Controls: Revision: 00
Design Control   Date: March 1, 1997
Supersedes: New
Section: 4.4
Page 1 of 3

4.4 Design Control

4.4.1 General

XYZ maintains procedures and instructions that define, control, and verify the design activities as specified in the requirements.

4.4.2 Design and Development Planning

The product manager is responsible for defining a product design, development plan, and implementation that describe the major activities of the project. It is also the product manager's responsibility to assign the design and verification activities to personnel having adequate capabilities and resources. A plan is created for each new product design or major redesign of existing products, and the plan is reviewed and approved before detailed design or development begins. The plan is also updated as the product evolves through the stages of design, development, and release for manufacturing (production).

It is the project manager's responsibility to make sure that all required skills are met on a per job basis following the R&M Guidelines.

4.4.3 Organization and Technical Interfaces

The XYZ company uses project managers in the engineering department for the deployment of new products. As such, the responsibility for organizational and technical interfaces in design control falls under the project manager's jurisdiction. Some of the responsibilities include, but are not limited to,

- Defining the appropriate channels for communication
- Encouraging proper documentation and review
- Defining the appropriate organizational interfaces
- Establishing appropriate cross-functional teams

4.4.4 Design Input

All design input requirements are identified, specified, communicated, and reviewed by the appropriate personnel within the XYZ company. During this review, the appropriate specification sheet is developed, and concerns and/or resolution of incomplete, ambiguous, or conflicting requirements by those responsible for their definition and specification are addressed and documented.

XYZ company uses computer aided design, engineering, and analysis in their design activities. The CAD/CAE systems are capable of two-way interfacing with customer systems.

Document No: QM XYZ 001
Title: XYZ's Quality Systems Practices and Controls: Revision: 00
        Design Control   Date: March 1, 1997
                                                                Supersedes: New
                                                                   Section: 4.4
                                                                  Page 2 of 3

4.4.5 Design Output

The vice president of engineering (or, if appropriately delegated, the product manager or project manager) is responsible for ensuring that design output is documented and expressed in terms of requirements, calculations, analysis, or other means that can be verified against design input requirements. Specific output includes analysis of test data and projections of reliability, maintainability, durability, and life cycle costs.

4.4.6 Design Review

XYZ conducts formal design reviews. They are planned and documented at the following six design stages according to the internal engineering procedures:

1. Concept
2. Definition
3. Design
4. Fabrication and assembly (prototype)
5. Testing/verification
6. Documentation and post-mortem review

Design reviews are coordinated by the product manager and/or the project manager and always include cross-functional teams concerned with the design stage being reviewed. Records of all reviews are maintained in the design project folder for a minimum of three years.

4.4.7 Design Verification

The vice president of engineering (or an assigned product manager or project manager) is responsible for ensuring that activities for verifying the design are defined, planned, executed, and documented by the appropriate personnel according to the prescribed procedures. Design verification will be conducted at least during the following stages prior to release:

• At the appropriate prescribed stage of design (see Clause 4.4.6).

• During performing product performance, maintainability, product life cost, life, reliability, durability, and other qualification tests as appropriate and tracking the testing to ensure timely completion and conformance to requirements.

Title: XYZ's Quality Systems Practices and Controls:
   Design Control

• Following a prototype development with predictive reliability and maintainability techniques.

• At the end of a specific analysis, comparisons with similar proven designs (if available), accelerated life testing as prescribed in the R&M Guidelines and other competitive or benchmark standards.

4.4.8 Design Validation

The vice president of engineering or an assigned project manager is responsible for ensuring that all final product meets and/or exceeds specifications prior to release for next-stage processing or shipping. Design validations are conducted and documented for the product after completion of a successful design verification.

4.4.9 Design Changes

All design changes and/or modifications are identified and transmitted through a Nonconforming Action Report (NCAR) or Engineering Services Request (ESR) to a master log through each phase of the machinery build. Design changes are initiated and coordinated by the product manager, product engineer, or project manager, and all changes/modifications are reviewed and approved by the authorized personnel and the customer (when required) prior to rerelease. The design changes are part of the document control system. In the case of changes to proprietary designs, the impacts on form, fit, function, performance, and durability are determined in consultation with the customer.

Related Documentation: (Place here all related documentation if appropriate and applicable.)

Title: XYZ's Quality Systems Practices and Controls:	Document No: QM XYZ 001
Document and Data Control	Revision: 00
	Date: March 1, 1997
	Supersedes: New
	Section: 4.5
	Page 1 of 1

4.5 Document and Data Control

4.5.1 General

XYZ company identifies and controls documents and data in all media that relate to the ISO/QS/TE requirements, as well as documents and data supplied by the customer or other sources and used to provide products and services. Whenever appropriate, special characteristics are identified.

4.5.2 Document and Data Approval and Issue

It is the responsibility of the quality assurance manager to ensure that all quality system documents (Quality Manual, Quality System Procedures, and the Quality System-related Work Instructions) have the following controls in place:

- All documents in the quality system are current.
- All control documents are properly signed.
- All revisions are properly signed and distributed.
- All obsolete documents have been removed from the system.
- All control documents have been reviewed appropriately and approved by the department manager prior to distribution and use.
- All documents and their copies are numbered and assigned to an individual or area of use.
- All control documents may be found in the master control document manual with the current names and locations of all holders.

XYZ has established procedures to ensure the timely review, distribution, and implementation of all customer engineering standards/specifications and changes.

4.5.3. Document and Data Changes

All document and data changes and modifications are made, reviewed, approved, identified, and communicated based on the following criteria:

- Document and data changes and modifications are reviewed and approved by the same personnel or departments that performed the original review and approval unless specifically designated otherwise.

- When applicable, the nature of the changes are identified in the document or the appropriate attachments according to the means specified in the related procedures for developing procedures and/or work instructions.

- XYZ uses the master control document manual to indicate the current revision of documents, and thereby prevent the circulation and use of obsolete documents (see Section 4.5.2).

Related Documentation: (Place here all related documentation if appropriate and applicable.)

Title: XYZ's Quality Systems Practices and Controls: Revision: 00
Purchasing   Date: March 1, 1997

4.6 Purchasing

4.6.1 General

The manager of purchasing is responsible for ensuring that all purchased products and subcontracted services from suppliers (including customer-designated vendors) that have an impact on the quality of XYZ's products or services conform to specified requirements. These control activities are conducted according to documented procedures as described below.

> • Materials for ongoing assembly of product for a customer with an approved subcontractor list are purchased only from suppliers on the list. Requests for additional suppliers must be submitted to the customer's materials engineering activity and approved prior to placing any orders.

> • All materials used in manufacturing must satisfy current governmental and safety constraints on restricted, toxic, and hazardous materials, as well as environmental, electrical, and electromagnetic considerations applicable to the country of manufacture and sale.

4.6.2 Evaluation of Subcontractors

XYZ requires all their subcontractors to follow prescribed quality procedures in providing parts and/or services. The qualification process is the responsibility of the purchasing manager. The following activities are the minimum:

> • Establishing an approved supplier list (ASL) on the basis of defined criteria related to a supplier's ability to meet XYZ's requirements for quality, cost, and delivery.

> • Maintaining the ASL based on supplier performance and reviews of supplier capability versus XYZ's requirements.

> • Scheduling and conducting supplier analyses/evaluations via a second- and or third-party audit against QS-9000 or ISO 9000.

> • Maintaining records of supplier capability, delivery, and performance

4.6.3 Purchasing Data

The purchasing manager is responsible for ensuring that purchase orders are reviewed and approved for adequacy of specified requirements.

Document No: QM XYZ 001
Revision: 00
Date: March 1, 1997
Supersedes: New
Section: 4.6
Page 2 of 2

Title: XYZ's Quality Systems Practices and Controls:
Purchasing

4.6.4 Verification of Purchased Product

4.6.4.1 XYZ's Verification at Supplier's Premises

> The manager of purchasing is responsible for ensuring that verification arrangements and the methods for product release are clearly defined in the purchasing documents for those situations when personnel from XYZ verify purchased product at the supplier's premises.

4.6.4.2 Customer Verification of Subcontracted Product

It is the policy at XYZ to allow the customer to inspect and verify that any subcontracted (vendored) product conforms to specified requirements, provided such verification is specified in the purchasing contract. Verification by the customer does not

> • Absolve XYZ of our responsibility to provide acceptable product.
>
> • Preclude the reality of, or the customer's right to, rejection of the product at a subsequent time.
>
> • Serve as evidence of effective quality control by the supplier.

Related Documentation: (Place here all related documentation if appropriate and applicable.)

Title: XYZ's Quality Systems Practices and Controls:
        Control of Customer-Supplied Product

4.7 Control of Customer-Supplied Product

XYZ has established and maintains current documented procedures for ensuring that all component parts, subassemblies, and test materials supplied by our customers are verified against specified requirements, identified, and maintained in adequate storage until required for use or incorporated into our products.

Related Documentation: (Place here all related documentation if appropriate and applicable.)

Title: XYZ's Quality Systems Practices and Controls:
Product Identification and Traceability

Document No: QM XYZ 001
Revision: 00
Date: March 1, 1997
Supersedes: New
Section: 4.8
Page 1 of 1

4.8 Product Identification and Traceability

XYZ establishes and maintains a documented tracking system (procedure) for identifying raw materials and supplies, component parts, subassemblies, and finished products by means of applicable drawings, specifications, and other documents from receipt and throughout all stages of production, delivery, and installation. Product identification is maintained and controlled. For incoming material,

- The material coordinator is responsible for ensuring that all incoming materials are clearly identified either individually (when appropriate) or as a lot while they are located at incoming inspection or in transit to storage or for use in production.

- The material coordinator maintains records identifying incoming materials by part number and their corresponding purchasing documentation such as specifications, inspection requirements, acceptance criteria, and other pertinent data.

- XYZ uses computer technology (when applicable) to record the receipt of materials, maintain accurate and timely records on inventory location and age, and update inventory status on a real-time basis.

For in-process material, the process supervisor is responsible for ensuring that in-process materials are clearly identified.

- All orders are "tagged" by means of a job/operation sheet that identifies the material (part number and description), shows its routing, and denotes its current state of processing.

- Floor stock items are identified by part number and part description.

- Stocked component parts are identified by part number and drawing revision number.

For finished products, the shipping supervisor is responsible for ensuring that finished products manufactured at XYZ are identified by means of a model number, serial number, and other pertinent data as required by the contract.

General Traceability: XYZ maintains records for each finished product that include the applicable specification sheets, bills of materials, engineering drawings, and contract specifications used for manufacture. Records on traceability are maintained for a period of five years. All traceability documents are controlled.

Related Documentation: (Place here all related documentation if appropriate and applicable.)

Document No: QM XYZ 001
Title: XYZ's Quality Systems Practices and Controls: Revision: 00
        Process Control   Date: March 1, 1997
                                                                    Supersedes: New
                                                                    Section: 4.9
                                                                    Page 1 of 3

4.9 Process Control

Managers of departments involved in assembly processes that directly affect quality
of intermediate and end products are responsible for ensuring that these processes
are identified, planned, and executed under controlled conditions. Controlled
conditions are defined to include the following requirements:

> • Documented procedures and/or work instructions for production,
> installation, and servicing, when their absence would adversely affect
> quality.
>
> • Suitable equipment and working environment, to include compliance
> with government regulations related to safety, environmental, and
> hazardous material.
>
> • Compliance with reference standards, codes, and quality plans and/
> or documented procedures.
>
> • Monitoring and control of suitable process and product characteris-
> tics, with an emphasis on those characteristics designated as "special"
> by the customer or by XYZ, during production, installation, and
> servicing. Special characteristics are designated, documented, and
> controlled as required by the customer.
>
> • Approval of processes and equipment, as appropriate.
>
> • Criteria for workmanship, which are either written or expressed by
> means of representative samples.
>
> • Equipment maintenance practices to ensure suitable and con-
> tinuing process capability, as described in the Preventive Mainte-
> nance Procedure and performed according to regularly updated
> and maintained schedules of maintenance activities for key pro-
> cess equipment. The maintenance program at XYZ includes predic-
> tive maintenance activities based on reviews of equipment-related
> information, such as manufacturers' recommendations, capability
> studies, and SPC results, tool wear, and fluid analysis, etc.

4.9.1 Process Monitoring and Operator Instructions

The quality assurance and/or the project manager and/or the product quality
planning team is responsible for ensuring that work instructions for processing are
developed from the sources given in the *Advanced Product Quality Planning and
Control Plan Reference Manual* and documented in the appropriate procedures.

Document No: QM XYZ 001
Revision: 00
Date: March 1, 1997
Supersedes: New
Section: 4.9
Page 2 of 3

Title: XYZ's Quality Systems Practices and Controls:
   Process Control

Furthermore, the manager of each department with process monitoring and operator instructions is responsible for ensuring that these work instructions are current, appropriate, understandable, sufficiently detailed, and accessible at the workstation, and that they are reviewed at the time of management reviews. All instructions will include the appropriate reference.

4.9.2 Preliminary Process Capability Requirements

Quality assurance and/or the process team of each individual process is responsible for ensuring that a preliminary process capability study plan that covers each of the special characteristics identified in the Control Plan is developed and reviewed with the customer, that the plan follows the customer's requirements and, if applicable, the *Fundamental Statistical Process Control* reference manual. Unless customer requirements dictate otherwise, a P_{pk} target of 1.67 is established for preliminary results (up to 30 days) and chronically unstable processes. For processes that produce attributes data, the data collected during preliminary production runs are used to set priorities for improvements and to establish control charting parameters based on the 50/20 run test.

If the results are an unacceptable preliminary capability, then the Corrective Action process (Section 4.14) and an evaluation of the mistake-proofing activities are initiated.

4.9.3 Ongoing Process Performance Requirements

At XYZ company, the goal of our quality is to continually improve and demonstrate that improvement. The established requirements for performance improvement (unless otherwise defined) are:

- C_{pk} target of at least 1.33 for stable processes and normally distributed data. For processes that demonstrate a consistently high C_{pk}, the control plan may be reviewed and revised to reflect the process performance—if the customer agrees.

- P_{pk} target of at least 1.67 for chronically unstable processes with output meeting specifications and a predicable pattern.

- Parts per million, nonparametric analysis, and/or index techniques to determine performance targets for nonnormal data.

- When these targets are not met and characteristic is found to be either unstable or noncapable, the reaction plan specified in the Control Plan and the Corrective Action Process (Section 4.14) are initiated to ensure that the procss is returned to a stable, capable state. Both the reaction plan from the Control Plan and the Corrective Action Plan are reviewed and approved by the customer when required.

Title: XYZ's Quality Systems Practices and Controls:
        Process Control

Document No: QM XYZ 001
Revision: 00
Date: March 1, 1997
Supersedes: New
Section: 4.8
Page 3 of 3

4.9.4 Modified Preliminary or Ongoing Capability Requirements

At XYZ, it is our policy that the Control Plan always reflects any customer-required modifications to the default capability requirements listed in the preceding sections.

4.9.5 Verification of Job Setups

Not Applicable.

4.9.6 Process Changes

Not Applicable.

4.9.7 Appearance Items

Not Applicable.

Related Documentation: (Place here all related documentation if appropriate and applicable.)

Document No: QM XYZ 001
Revision: 00
Date: March 1, 1997
Supersedes: New
Section: 4.10
Page 1 of 2

Title: XYZ's Quality Systems Practices and Controls:
Inspection and Testing

4.10 Inspection and Testing

4.10.1 General

XYZ has established and properly maintains documented procedures and/or work instructions or quality plans that define the required inspection and testing activities and related records used to verify that all product requirements are met prior to product distribution, processing, or use.

The acceptance criteria are based on the notion of "zero defects"; however, the specific requirement(s) of the customer prevail. All criteria are identified in the Control Plan. When certification is necessary from laboratory facilities, that is also identified in the Control Plan.

4.10.2 Receiving Inspection and Testing

In most cases, most of the receiving inspection and testing is waived due to either third-party certification and/or statistical documentation at the time of delivery. Appropriate documentation is available. When neither is verifiable, then

- The specific processes, both manufacturing (production) and non-manufacturing (nonproduction), are responsible for ensuring that an incoming product and/or service is not used or processed until it has been verified as appropriate and conforming to the requirements.

- The amount and nature of receiving inspection and testing depends on the requirements of the customer, the history of the supplier, and the supplier's ability to supply appropriate documentation.

- All verification is done through procedures and specific instructions.

- All appropriate documentation, usage, and storage of inspection records are maintained and available to appropriate personnel.

- Appropriate segregation of nonconforming material is identified and mandatory.

- When an emergency arises and the material is needed for an urgent production situation, that specific material is identified in a manner to allow its recall and replacement in the event the product is determined at a later time not to be in compliance with specified requirements. This special procedure is also addressed with appropriate procedures.

Document No: QM XYZ 001

Title: XYZ's Quality Systems Practices and Controls: Revision: 00
      Inspection and Testing   Date: March 1, 1997
                                                                  Supersedes: New
                                                                  Section: 4.10
                                                                  Page 2 of 2

4.10.3 In-Process Inspection and Testing

Quality assurance and/or project management has the responsibility for ensuring that in-process product is held and not used or processed further until it has been inspected, tested, or otherwise verified as conforming to specified requirements, except when product is released under positive recall (see Section 10.2). It is XYZ's policy to direct process activities and resources toward defect prevention rather than relying simply on traditional methods of defect detection.

4.10.4 Final Inspection and Testing

The compliance and/or the quality assurance department have the responsibility for ensuring that no product is dispatched until (a) all final inspection and testing are complete according to the appropriate documented procedure and/or work instruction or quality plan to show evidence of (product) conformance to specified requirements; (b) all data and documentation covering all inspections and tests (incoming, in process, and final) specified in the quality procedures and control plans are available and authorized to show compliance, and the results meet specified requirements; and (c) all functional verification has been performed based on customer and qualification runoff requirements.

Appropriate and applicable documentation, usage, and maintenance of records are maintained and available.

4.10.5 Inspection and Test Records

XYZ's inspection and test records are established and maintained to identify the persons performing inspection and test activities and the results of these verification activities. The responsibility for generating, filing, and maintaining inspection and test records is defined in the appropriate procedures.

Related Documentation: (Place here all related documentation if appropriate and applicable.)

Document No: QM XYZ 001
Revision: 00
Date: March 1, 1997
Supersedes: New
Section: 4.11
Page 1 of 2

Title: XYZ's Quality Systems Practices and Controls:
Control of Inspection, Measuring, and
Test Equipment

4.1 Control of Inspection, Measuring, and Test Equipment

4.11.1 General

The manager of quality assurance is responsible for establishing and maintaining
documented procedures and work instructions for ensuring that all inspection,
measuring, and test equipment used in any stage of production or installation is
controlled, calibrated, and properly maintained to demonstrate the conformance of
product to the specified requirements.

When measurement is necessary, the appropriate test equipment and calibration are
accounted for with appropriate and applicable documentation. To facilitate this,
XYZ company provides available technical data regarding measurement device
calibration to its customers (on request) to allow customer verification that the
measurement devices are functionally adequate.

4.11.2 Control Procedure

The XYZ company has formal, documented procedure and work instructions for
regularly certifying the accuracy of every inspection instrument, including instruments owned by employees, used to make quality decisions in the manufacturing
process. Appropriate procedures and instructions are established, are available to
authorized personnel, and maintained.

All inspection, measuring, and test equipment used to verify dimensions or
characteristics, or perform functional testing and thereby accept parts, subassemblies, or assemblies, are calibrated periodically, with all appropriate documentation
maintained and available.

A calibration checklist, the record of calibration, and all activities (including outside
sources) related to calibration are properly maintained and available. The procedures and instructions for such activities may be found in the appropriate manuals.

All internal standards that are used to verify the accuracy of inspection instruments
or assembly tools are regularly calibrated by outside labs. Attribute-type gages are
sent to outside labs to be cleaned, repaired, and/or calibrated. All calibration
services, whether performed in or outside the plant, are certified to use standards
that are traceable to the National Institute of Standards and Technology (NIST).

All inspection, measuring, and test equipment that are not in current calibration are
not used. New equipment or equipment with a past due calibration date are
impounded to prevent use until the calibration has been completed.

Document No: QM XYZ 001

Title: XYZ's Quality Systems Practices and Controls: Revision: 00
    Control of Inspection, Measuring and   Date: March 1, 1997
    Test Equipment   Supersedes: New
                      Section: 4.11
                      Page 2 of 2

All XYZ employees that use the measuring and test equipment are responsible for checking the calibration sticker (or equivalent tag/marker) to ensure that the calibration status is current. Should calibration activity disclose the potential for discrepant material that has been shipped, the customer shall be notified of all pertinent information. This notification may be followed by a request for waiver.

Calibration activity that discloses the potential for nonconforming material that is still within the facilities of XYZ results in the initiation of an ad hoc audit for the purpose of determining whether or not the potential was realized. If the material is found to be nonconforming, the appropriate procedure is initiated; otherwise, the material continues its normal production course.

4.11.3 Inspection, Measuring, and Test Equipment Records

All XYZ's inspection, measuring, and test equipment records are generated and controlled as stated in Section 4.11.2. The manager of quality assurance is responsible for the maintenance of all calibration records, including communications to customers regarding the potential for nonconforming product due to out-of-calibration equipment, and for ensuring that the control of measurement equipment meets or exceeds the requirements stated in the *Measurement Systems Analysis* reference manual.

4.11.4 Measurement System Analysis

The manager of quality assurance is responsible for maintaining records of gage repeatability and reproducability (GR&R) and other applicable statistical studies to analyze measurement system uncertainty. XYZ has documented work instructions that include analytical methods and acceptance criteria that meet the requirements specified in the *Measurement Systems Analysis* reference manual.

Related Documentation: (Place here all related documentation if appropriate and applicable.)

Document No: QM XYZ 001

Title: XYZ's Quality Systems Practices and Controls: Revision: 00
    Inspection and Test Status   Date: March 1, 1997
                                                                  Supersedes: New
                                                                  Section: 4.12
                                                                  Page 1 of 1

4.12 Inspection and Test Status

XYZ has a system that identifies the inspection and test status of all products by
use of markings, authorized stamps, tags, labels, routing cards, inspection records,
physical location designations, or other suitable means; these indicate the conformance or nonconformance of the product with regard to the inspections or tests
performed. The identification of inspection and test status is maintained, as
defined in the company's procedures and work instructions and as required by
customers, throughout production and installation of the product to ensure that
only product that has passed the required inspections and tests (or has been
released under an authorized concession) is dispatched, used, or installed.

The responsibility to identify the inspection and test status of products lies with the
manager of the department responsible for performing the inspection and/or test.
These responsibilities are formally stated in the applicable inspection and testing
procedures and/or work instructions.

Related Documentation: (Place here all related documentation if appropriate and
applicable.)

Title: XYZ's Quality Systems Practices and Controls:
    Control of Nonconforming Product

4.13 Control of Nonconforming Product

4.13.1 General

The manager of quality assurance is responsible for maintaining documented procedures and work instructions for ensuring that product not conforming to specified requirements is clearly identified and quarantined or segregated to prevent inadvertent use or installation until the material is reviewed and disposition is determined. The manager of quality assurance is also responsible for maintaining and analyzing data from the nonconforming material.

4.13.2 Review and Disposition of Nonconforming Product

XYZ's review policies for nonconforming product are followed by all personnel who detect suspect or nonconforming material. In fact, XYZ's policy is that the person producing the product is also responsible for its quality.

When products are found to be nonconforming, they are immediately tagged with a "Do Not Use" tag. The product is physically segregated (whenever possible) or is prevented from further processing or use by other means. At that point, appropriate personnel are identified, and proper action based on the specified procedures and instructions takes place.

A product that is reworked or repaired is reinspected according to XYZ's appropriate inspection and test procedure and/or control plan. Pertinent instructions for rework are readily available and used by all appropriate personnel.

4.13.3 Control of Reworked Product

All rework is controlled by the appropriate and applicable procedures and/or instructions either to make the product usable or to scrap it.

4.13.4 Engineering-Approved Product Authorization

Not Applicable.

Related Documentation: (Place here all related documentation if appropriate and applicable.)

Document No: QM XYZ 001

Title: XYZ's Quality Systems Practices and Controls: Revision: 00
    Corrective and Preventive Action   Date: March 1, 1997

Supersedes: New

Section: 4.14

Page 1 of 2

4.14 Corrective and Preventive Action

4.14.1 General

XYZ's focus on quality is prevention. As such, the company has established and
currently maintains documented procedures and related records for implementing
both corrective and preventive action. These procedures specify problem-solving
actions for eliminating the cause of actual or potential quality system problems and
related nonconformities to a degree commensurate with the magnitude of the
problem, its potential outcome, and the level of risk involved. When resolving
external nonconformities, corrective action shall take place as required by the
customer. The management representative for both the corrective and preventive
action(s) has been designated as the manager of quality assurance.

4.14.2 Corrective Action

XYZ's corrective action policy is to identify, analyze, and resolve problems that
have been raised by nonconformances. The result of the corrective action is to
correct the "root cause" of the problem and, when necessary, to revise the
company's quality system, procedures, and/or instructions, as appropriate.

XYZ's corrective action follows, as a general rule, the 8-D approach (other approaches are used as necessary) to resolve the problems identified either internally
or externally. All corrective actions are initiated, controlled, and documented
through the corrective action request form. Their retention is for three years unless
otherwise specified by the customer and/or governmental regulations. For the
appropriate reviews and documentation, see the relevant procedures and
instructions.

The responsible person for the corrective action is the quality manager.

4.14.3 Prevention Action

XYZ's focus on prevention is the basic tenet of quality. Everything we do is founded
on the principles of prevention. We look for weaknesses in the quality system and
capitalize on these opportunities to continually improve based on the findings of
internal, external, and/or third-party audits.

Document No: QM XYZ 001

Title: XYZ's Quality Systems Practices and Controls: Revision: 00
    Corrective and Preventive Action   Date: March 1, 1997
Supersedes: New
Section: 4.14
Page 2 of 2

Appropriate reviews of the use of information, quality records, service reports, customer feedback, special studies, and so on are conducted to identify and/or eliminate causes of potential nonconformances. The responsibility for prevention is with all employees; however, the manager of quality assurance is responsible for ensuring that the prevention program in XYZ is managed effectively.

Related Documentation: (Place here all related documentation if appropriate and applicable.)

Title: XYZ's Quality Systems Practices and Controls:
    Handling, Storage, Packaging, Preservation,
    and Delivery

Document No: QM XYZ 001
Revision: 00
Date: March 1, 1997
Supersedes: New
Section: 4.15
Page 1 of 2

4.15 Handling, Storage, Packaging Preservation, and Delivery

4.15.1 General

At XYZ, we believe that the managers responsible for handling, storage, packaging, and delivering materials and products are also responsible for establishing, documenting, and maintaining methods appropriate to satisfy the requirements in manufacturing and those specified by contract. The procedures and instructions relevant to these requirements are documented and available in the appropriate manuals and locations.

4.15.2 Handling

XYZ's policy for handling is to use methods and means appropriate for handling and transporting product in a manner that prevents loss of product value and satisfies customer and governmental requirements.

4.15.3 Storage

XYZ's policy for storage is to ensure the appropriate management of stored material according to the customer's requirements, governmental regulations, and our internal procedures and instructions. The person responsible for the appropriate storage is the plant manager. The focus of our policy is to control inventory turns, assure stock rotation, appropriately separate "good" from "not usable" material, and continuously maintain minimum inventory levels and prevention of their damage or deterioration.

4.15.4 Packaging

XYZ's policy on packaging is that all products are appropriately packed and identified on the packaging in a manner that allows for ready identification and traceability through all stages of processing and prevents the loss of product value. Customer packaging and labeling requirements are detailed in appropriate contract-related work instructions and guidelines, including the AIAG Shipping/Parts Identification Label Standard (AIAG B-3).

Document No: QM XYZ 001

Title: XYZ's Quality Systems Practices and Controls: Revision: 00
       Handling, Storage, Packaging, Preservation,   Date: March 1, 1997
       and Delivery   Supersedes: New
                                                                    Section: 4.15
                                                                    Page 2 of 2

4.15.5 Preservation

At XYZ, all products (incoming and in process) are segregated and preserved as necessary to maintain product quality and value through all stages under the company's control.

4.15.6 Delivery

At XYZ, the vice president of manufacturing (and, quite often, the plant manager with the aid of the quality manager) is responsible for ensuring that the quality of the final product is protected after final inspection and test according to the appropriate procedure. When contractually specified, XYZ company shall be responsible for packaging and preservation during transit, including delivery to its destination.

Delivery performance is one of the key indicators for both the XYZ company and our customers. As such, we track on-time deliveries, production scheduling, and all advanced shipment notifications. When we discover problems and/or concerns, we initiate a corrective action report, and the results of the investigation are communicated to both our management and customer(s).

To make sure delivery meets the customer's expectations, appropriate timing plans, resource plans, and scheduling systems are established throughout the XYZ company.

Subcontractors are also monitored for appropriate scheduling systems.

Related Documentation: (Place here all related documentation if appropriate and applicable.)

Document No: QM XYZ 001

Title: XYZ's Quality Systems Practices and Controls: Revision: 00
        Control of Quality Records   Date: March 1, 1997
                                                                        Supersedes: New
                                                                        Section: 4.16
                                                                        Page 1 of 1

4.16 Control of Quality Records

XYZ's quality system is documented through the use of quality records. Quality records may be those records that XYZ generates and/or customers and suppliers provide.

The responsibility for establishing, collecting, filing, indexing, storing, maintaining, and disposing of records is defined in XYZ's quality-system-related documentation. The retention of the quality records must be maintained for at least three years unless the customer and/or governmental regulations dictate otherwise.

Related Documentation: (Place here all related documentation if appropriate and applicable.)

Document No: QM XYZ 001

Title: XYZ's Quality Systems Practices and Controls: Revision: 00
       Internal Quality Audits   Date: March 1, 1997
                                                          Supersedes: New
                                                            Section: 4.17
                                                             Page 1 of 1

4.17 Internal Quality Audits

XYZ plans and conducts at least two internal quality audits annually according to
the ISO/QS and customer requirements. The results of the audit are documented and
communicated to management and all other appropriate personnel for any appropriate action. The focus of our audits is to verify the effectiveness of our quality
system and to discover opportunities for improvement.

The quality assurance manager is responsible for organizing and coordinating the
internal audit, as well as maintaining appropriate records for such audits.

Related Documentation: (Place here all related documentation if appropriate and
applicable.)

Title: XYZ's Quality Systems Practices and Controls:
     Training

Document No: QM XYZ 001
Revision: 00
Date: March 1, 1997
Supersedes: New
Section: 4.18
Page 1 of 1

4.18 Training

At XYZ, people are the company's most valuable asset. Investing in people (both
management and nonmanagement) through effective training is thus a key corporate
strategy for achieving the company's mission and quality policy. Therefore, it is our
policy to identify the training needs and provide training of all personnel performing
activities affecting quality. Documentation for all the training is maintained and
may be found in the personnel department.

The effectiveness of the training is established through needs assesment for the
appropriate training and then through course evaluations, follow-up surveys, and
documented improvements in the affected areas.

At XYZ, we value training so highly that all company managers have a contributing responsibility in assessing training needs, providing on-the-job reinforce-
ment of skills, and evaluating the effectiveness of training given to the personnel they directly manage. Furthermore, to reinforce this responsibility, we have
empowered our employees to request training at any time if the employee feels
that training is essential for providing knowledge and skills required to
maintain the requirements of the standard and the employee's job responsibility.

Special effort throughout the XYZ company is being made to train all appropriate
personnel in reliability and maintainability methods.

Related Documentation: (Place here all related documentation if appropriate and
applicable.)

Document No: QM XYZ 001

Title: XYZ's Quality Systems Practices and Controls: Revision: 00
    Servicing   Date: March 1, 1997
    Supersedes: New
    Section: 4.19
    Page 1 of 1

4.19 Servicing

XYZ maintains documented procedures for providing contracted services that meet specified requirements and yield high levels of customer satisfaction and for reporting on the results of such services to appropriate activities throughout the organization. Servicing at XYZ takes the following forms:

- Managing customer interfacing

- Managing customer complaints

- Providing service parts and appropriate training

- Providing field service on a contractual basis when applicable

The responsibility of servicing belongs to the marketing department (except when otherwise identified, e.g., quality issues belong to the quality department). The sales manager and/or quality manager—as necessary—will coordinate the customer service activities and ensure that appropriate records to document customer service performance are maintained.

Special effort by the marketing department is made to communicate to the customer and all appropriate personnel information on machine uptime, reliability, maintainability, maintenance history, and service concerns.

Related Documentation: (Place here all related documentation if appropriate and applicable.)

Title: XYZ's Quality Systems Practices and Controls:
      Statistical Techniques

4.20 Statistical Techniques

4.20.1 Identification of Need

XYZ recognizes that statistical techniques are valuable for assessing, controlling, and improving our quality system and processes. The concepts of variation, control (stability), capability, and overadjustment are communicated and understood throughout our company. For our processes, we utilize both qualitative and quantative data. Therefore, the specific need and application for statistical techniques and the establishment of specific methods and instructions for improvement of a specific process are assumed by all area managers with the collaboration of the quality manager and/or quality engineer (the resolution of the appropriate tool is established through quality planning). Once the selection of the appropriate statistical tools for individual processes is established, it is documented in the Control Plan.

4.20.2 Applications and Procedures

XYZ uses both statistical and nonstatistical methods to maintain process control, monitor defect prevention, assess machine capabilities and levels of quality, and identify areas for quality improvement. Examples of some of the specific tools that our company utilizes are the following:

- brainstorming
- process flow charts
- histograms
- Pareto charts
- scatter plots
- force field analysis
- variable and attribute charts
- design of experiments (classical, Taguchi)
- FMEA
- Gage R&R
- MTBF
- MTTR
- short-run SPC

The specific application of these tools depends on the individual process, and the reader of this manual is encouraged to see the individual process for further information.

Related Documentation: (Place here all related documentation if appropriate and applicable.)

Title: XYZ's Quality Systems Practices and Controls: Revision: 00
        Sector Specific Requirements—Purpose   Date: March 1, 1997
                                                                Supersedes: New
                                                                Section: II: 1
                                                                Page 1 of 1

Section II. 1.0 Purpose

XYZ documents, reviews, submits, revises, and maintains records for all its products. Appropriate and applicable documentation exists to support customer satisfaction, reliability, and functional performance prior to the machinery being integrated into the customer's plant.

Section II. 1.1 Machinery Qualification Runoff Requirements

The vice president of manufacturing is responsible for ensuring that all changes are suitably validated according to the guidelines provided in the machinery qualification runoff requirements.

Section II. 1.2 Procedure

At XYZ, we conduct appropriate and applicable testing to meet the 20-hour run testing.

Related Documentation: (Place here all related documentation if appropriate and applicable.)

	Document No: QM XYZ 001
Title: XYZ's Quality Systems Practices and Controls:	Revision: 00
Sector Specific Requirements—	Date: March 1, 1997
Continuous Improvement	Supersedes: New
	Section: II: 2
	Page 1 of 1

Section II. 2.0 Continuous Improvement

Section II. 2.1 General

It is the joint responsibility of the management representative and the manager of quality assurance to ensure that the concepts of continuous improvement and the principles of ISO/QS are communicated throughout the XYZ company and that personnel receive the appropriate education and training in continuous improvement tools and techniques.

It is the policy of XYZ company to strive to continuously improve the areas of quality, service, delivery, and price for all customers, as evidenced by and tracked in the appropriate documentation and customer satisfaction indices.

Action plans for the improvement of key processes in production, business, and support functions are developed and documented in the Business Plan in a continuous effort to reduce variation and increase productivity and efficiency. Special emphasis is given to those processes identified as most important to XYZ's customers.

Section II. 2.2 Quality and Productivity Improvements

Improvement projects and plans are documented in the Business Plan and are developed based on the indicators tracked in the appropriate documentation. Appropriate measurables for the improvement projects are defined during project planning and are tracked and revised as appropriate throughout the life of the project.

Section II. 2.3 Techniques for Continuous Improvement

The XYZ management representative and manager of quality assurance are jointly responsible for ensuring that all personnel understand and apply the principles of continuous improvement. In addition, they are also responsible for making sure that appropriate education and training is given to the appropriate employees in advanced methods and techniques for continuous improvement, as appropriate. The tools and methods used at XYZ company meet, and in most cases exceed, the requirements of the standards.

Related Documentation: (Place here all related documentation if appropriate and applicable.)

Title: XYZ's Quality Systems Practices and Controls:
    Sector Specific Requirements—
    Manufacturing Capabilities

Section II. 3.0. Manufacturing Capabilities

Section II. 3.1 Facilities, Equipment, and Process Planning and Effectiveness

XYZ's objective is to optimize process and material flows, floor plans, and use of floor space. To meet this requirement, XYZ has identified the product quality planning team as the responsible party for evaluating the effectiveness of the manufacturing system during the advanced quality planning process and for ensuring that a cross-functional team approach is used in developing the plans for facilities, equipment, and processes.

The specific responsibility for the evaluation of effectiveness belongs to the vice president of manufacturing, who ensures the implementation, documentation, and review of regular assessments of the manufacturing system. Some factors for the evaluation are plant layout, work flow, automation and human factors issues, planned and predictive maintenance programs, balancing of operators and lines, storage and buffer inventory levels, and value-added labor content.

Section II. 3.2 Mistake-Proofing

XYZ company is committed to devoting resources to problem prevention rather than relying simply on problem detection. Therefore, mistake-proofing activities take place throughout the planning process, as well as during problem resolution. The results of design and process FMEAs, reliability studies, the use of CAD systems, capability studies, the preventive action process, and service reports are used to identify areas in which to apply mistake-proofing methodologies.

Section II. 3.3 Tool Design and Fabrication

The manager of design engineering is responsible for establishing and implementing documented procedures and work instructions for tool and gage design, fabrication, and complete dimensional inspection. This manager is also responsible for ensuring that appropriate technical resources and facilities exist for these activities. Customer-owned tools and gages are permanently identified as such.

Section II. 3.4 Tooling Management

The plant manager is responsible for establishing and implementing documented procedures and work instructions for tool and gage maintenance, repair, storage, recovery, setup, and replacement. The plant manager is also responsible for ensuring that appropriate technical resources and facilities exist for these activities.

Related Documentation: (Place here all related documentation if appropriate and applicable.)

Document No: QM XYZ 001

Title: XYZ's Quality Systems Practices and Controls: Revision: 00
    Customer-Specific Requirements   Date: March 1, 1997
                                                                      Supersedes: New
                                                                        Section: III
                                                                       Page 1 of 2

Section III. Customer-Specific Requirements

> Note: In this section of the manual, you should include the specific requirements
> that Chrysler Corporation, Ford Motor Corporation, and General Motors
> have defined as important and are not covered as part of Section I or II.

This section of the manual is customized for each company. Because of this high
customization level, we cannot provide the reader with specific examples, other than
very generic guidelines: make sure before you embark on writing or modifying your
quality manual, that you come in contact with your customer's quality supplier
representative and both of you agree as to what should be included in this section.
In any case, however, typical content for this section is shown below.

Section III.1 Chrysler Requirements

Unique requirements are
- Chrysler requires the supplier to have a system that identifies (a) safety
characteristics, (b) special characteristics, (c) critical tooling characteristics, and
(d) significant characteristics.
- Another requirement is to demonstrate continuous conformance by conducting
an annual layout inspection.
- Process performance P_{pk}.

Redundant requirements are
- Internal quality audits (see ISO 9002: Section 4.17)
- Design validation/product verification (see ISO 9001: Section 4.4)
- Corrective action plan (see ISO 9001: Section 4.14)
- Packaging, shipping, and labeling (see ISO 9001: Section 4.15)
- Process sign-off (see ISO 9001: Section 4.9)

Section III.2. Ford Requirements

Unique requirements are
- Appropriate control item parts and critical characteristics with the
inverted delta
- Control plans and failure mode and effect analyses (FMEAs)
- Design strategies, reviews
- Reliability and maintainability plan, testing, and assessment
- R&M continuous improvement activities
- Equipment and standard parts
- Heat treating
- Quality operating system (QOS)

Document No: QM XYZ 001
Title: XYZ's Quality Systems Practices and Controls: Revision: 00
    Customer-Specific Requirements   Date: March 1, 1997
                                                    Supersedes: New
                                                    Section: III
                                                    Page 2 of 2

Redundant requirements are
- Setup verification (see ISO-9001: Sections 4.4 and 4.9)
- Process changes and monitoring (see ISO 9001: Section 4.9)
- Engineering specification, prototypes, and qualifications for materials (see ISO 9000: Section 4.4)
- Labeling (see ISO 9001: Section 4.15)

Section III.3 General Motors' Requirements

Unique requirements are
- Technology program
- Key characteristic designation system
- Continuous improvement
- Run at rate
- Specifications for part and bar codes
- FMEA and control plans
- Design reviews
- Layout inspections
- Maintainability requirements
- Reliability and maintainability validation

Redundant requirements are
- Process approval (see ISO 9001: Section 4.9)
- Match check material (see ISO 9001: Sections 4.6 and 4.9)
- Traceability (see ISO 9001: Section 4.8)
- Problem reporting (see ISO 9001: Section 4.14)
- Evaluation of supplier (see ISO 9001: Section 4.6)
- Production containment procedure (see ISO 9001: Section 4.9)
- Packaging, shipping, and delivery and labeling (see ISO 9001: Section 4.15)
- Prototyping (see ISO 9001: Section 4.4)

Appendix B: Typical quality procedures and instructions

This appendix provides the reader with many samples of procedures and instructions specific to the tooling and equipment industry. They may serve as a model for generating your own organization's second, third, or even fourth level of documentation. As they are presented, they provide a model for structure and a generic content for a given company. Therefore, the reader should not use them as they are. With some specific organizational modifications to reflect one's quality system and approach, these procedures and instructions may indeed provide a starting point for full development of a complete documentation system.

Furthermore, the reader is warned that in some specific situations (e.g., deviation from specification, waiver of change, first article of inspection, etc.), individual companies may require the supplier to use their own forms.

SUBJECT: QUALITY PROCEDURES		
Authorized By: *V. A. Philis*		**Number:** 2.1.1
Effective Date: 2/10/97	**Revision:** 1/30/97	**Page** 1 **of** 3

1.0 PURPOSE

This procedure provides instructions for the uniform preparation and/or revision of quality procedures at the XYZ company, including the proper methods for securing appropriate approvals and distributed released copies.

2.0 APPLICATION

This procedure applies to all employees of the XYZ company involved in recommending the initiation or revision of quality procedures, as well as those involved in preparing or revising them, those authorized to approve them, and those controlling the distribution of approved copies.

3.0 DEFINITIONS

3.1   Quality procedures are the directives issued by quality assurance for communicating the established methods for performing and administering the work relative to assuring and controlling the quality of the company's products.

3.2   Quality procedures provide the summary-level information required on a given subject. If this information must be described in further detail for a specific application, this detail is to be recorded on work instructions (see Quality Procedures X.Y.P).

3.3   Quality procedures are to be formatted as follows, whenever applicable.
PURPOSE
APPLICATION
DEFINITIONS
ASSOCIATED MATERIALS
PROCEDURE

Under "PURPOSE" should be a clear statement as to the procedure's intent—its objective. Under "APPLICATION" should be a description of organization, functions, or items affected by the procedure. Under "DEFINITIONS" can be explanations of unique or special words or terms appropriate to the procedure. Under "ASSOCIATED MATERIALS" can be identification of specific documents or other materials associated with the procedure. Under "PROCEDURE" should be a clear description of the steps to be taken in accomplishing what is required.

All procedures should use the categories of PURPOSE, APPLICATION, and PROCEDURE. The categories of DEFINITIONS and ASSOCIATED MATERIALS can be used if they apply.

SUBJECT: QUALITY PROCEDURES		
Authorized By: *V. A. Philis*		**Number:** 2.1.1
Effective Date: 2/10/97	**Revision:** 1/30/97	**Page** 2 **of** 3

3.4   When rough and final drafts of new or revised procedures have been completed, they are to be distributed to departments affected by the procedure for coordination and approval purposes.

Each draft of the procedure is to be distributed using the Quality Procedure Coordination Cover Sheet shown in Figure B-1. After review and, if necessary, corrections, the form sheet is to be signed by the appropriate department manager, signifying approval of the procedure "as is" or "as corrected."

3.5   After all appropriate departmental approvals of the procedure's final draft have been secured, it is to be submitted to the vice president of administration for approval. The approval signature is to be placed on the remarks.

3.6   Approved quality procedures are to be distributed to Quality Assurance Manual holders.

3.7   All quality procedures are to be reviewed annually by the quality department of the XYZ company to ensure continued correctness and applicability.

4.0 ASSOCIATED MATERIALS

4.1   Quality Procedures Status Log — Preparation and Coordination, Form No. B-1 (top of form)

4.2   Quality Procedure Status Log —Approval and Requisition Form No. B-1 (bottom of form)

5.0 PROCEDURE

5.1   New quality procedures may be requested by any company employee. A rough draft should be prepared that outlines the purpose and procedural content. It should be submitted to the quality department, or dropped in the suggestion box.

5.2   Revisions to existing quality procedures may be requested by any company employee. The existing procedure should be clearly marked as to the recommended change. It should be submitted to the quality department.

5.3   Feedback to the person suggesting the change will be provided by the quality department personnel.

5.4   Quality procedures will be typewritten.

SUBJECT: QUALITY PROCEDURES		
Authorized By: *V. A. Philis*		**Number:** 2.1.1
Effective Date: 2/10/97	**Revision:** 1/30/97	**Page** 3 **of** 3

Figure B-1. Quality Procedure—Status log

Procedure Name:	Number:
Status (Check one) Rough Draft_____	Final Draft _____
Appproval required from department:	

If changes or corrections are required, please mark attached copy or comment here.

Approved as corrected:

Name/Title:	Department:	Date

Approved as is:

Name/Title:	Department:	Date

SUBJECT: QUALITY BULLETINS		
Authorized By: *V. A. Philis*		**Number:** 2.1.2
Effective Date: 2/10/97	**Revision:** 1/30/97	**Page** 1 **of** 3

1.0 PURPOSE

The purpose of this procedure is to describe the preparation and distribution of the quality bulletins used by the XYZ company.

2.0 APPLICATION

Quality bulletins are used to provide (1) emergency changes to existing quality procedures; (2) expedient interim applications of new, pending quality procedures; and (3) one-time or limited-use instructions or information relative to the quality assurance program.

3.0 ASSOCIATED MATERIALS

3.1   Quality Bulletin, Form No. B-2

4.0 PROCEDURE

4.1   Requests for use of the Quality Bulletin system should be submitted to the Quality Department.

4.2   If distribution is appropriate beyond the Quality Assurance Manual holders, a recommended distribution list should be submitted by the requester to the quality department.

4.3   The Quality Bulletin Form, as shown in Figure B-2, is to be prepared by the person who needs to request a change as follows: (1) the control number of the bulletin to be assigned by the quality assurance department; (2) the bulletin's date of release; and (3) the page number of the bulletin, starting with "1." If the bulletin supersedes a previously issued bulletin, the old number and date should be cited in (4) and (5). The subject should next be clearly given (6), followed by the complete text of the bulletin (7). Next (8), the form should be check-marked and date filled in, where applicable, relative to one of the four statements:

1. Retain this bulletin until further notice.

2. Discard this bulletin after noting contents.

3. This bulletin will be invalid after _____(date)_____.

SUBJECT: QUALITY BULLETINS		
Authorized By: *V. A. Philis*		Number: 2.1.2
Effective Date: 2/10/97	Revision: 1/30/97	Page 2 of 3

   4. This bulletin will be incorporated into Quality Procedure No. _____
by ___(date)___.

   When the information on the form is complete, it should be submitted to the manager of quality assurance for an approval signature (9). The signature of the requester is item (10). If the bulletin's instructions or information crosses department lines, other appropriate signatures may be required.

   4.4   When the bulletin has been signed, it is to be reproduced and distributed. Distribution is to be to all Quality Assurance Manual holders, plus (if provided) the additional names cited on the distribution list (Paragraph 4.2, above).

   4.5   Changes that require one week or longer to resolve may utilize a quality bulletin.

   4.6   The original will be maintained and kept in a file by the quality department.

SUBJECT: QUALITY BULLETINS			
Authorized By: *V. A. Philis*		**Number:** 2.1.2	
Effective Date: 2/10/97	**Revision:** 1/30/97	**Page** 3 **of** 3	

Figure B-2. Quality Bulletin

Quality Bulletin

Number _____ (1)
Date _____ (2)
Page _____ of ____ (3)
Supersedes No._____ (4)
Dated _____ (5)

Subject:

(6)

(7)

(9)
Manager of Quality Assurance

(8)

Retain this bulletin until further notice _____
Discard this bulletin after noting contents _____
This bulletin will be invalid after (data) _____
This bulletin will be incorported into quality
procedure No._____ by (date) _____

(10)
Signature of the requester

SUBJECT: QUALITY ASSURANCE AUDITS OF PRODUCTS AND SYSTEMS		
Authorized By: *V. A. Philis*		**Number:** 2.2.1
Effective Date: 2/10/97	**Revision:** 1/30/97	**Page** 1 **of** 3

1.0 PURPOSE

This procedure sets forth the method for periodically and randomly examining products and systems to determine the effectiveness of the overall quality assurance program.

2.0 APPLICATION

This procedure applies to the following internal functions: receiving, receiving inspection, commercial and standard stock, manufacturing, inspection, engineering, shipping, and other areas that affect product quality. The quality auditing of supplier and subcontractor activities is covered by another procedure (see Quality Procedure 2.3.1, Quality Assurance Audits of Subcontractors and Suppliers).

3.0 DEFINITIONS

3.1   Quality audit: An official examination of functions and systems that takes place on a periodic but random, unannounced basis to verify the effectiveness of the company's quality program. This should be done at least quarterly.

4.0 ASSOCIATED MATERIALS

4.1   Quality Assurance Audit Plan and Report, Form Nos. B-3.

4.2   Corrective Action Process, Quality Procedure (QP) 3.5.1

4.3 PROCEDURE

4.4   The quality management and administration department has the responsibility for planning and conducting quality audits.

4.5   An audit plan is first prepared using the Quality Assurance Audit Plan and Report Form (see Figure B-3). The plan section of this form is filled out as follows: (1) the assigned number of the audit plan and report, (2) the name of the function/subject, (3) the planned audit date, (4) the audit subject/department, (5) a detailed description of the audit plan, and (6) the signature of the person who has prepared the plan.

4.6   Audit department supervision reviews the plan and, following corrections, if any, approves the plan. Approval is signified by (7) signing and dating the form.

4.7   The audit now takes place per the detail plan.

SUBJECT: QUALITY ASSURANCE AUDITS OF PRODUCTS AND SYSTEMS			
Authorized By: *V. A. Philis*		Number: 2.2.1	
Effective Date: 2/10/97	Revision: 1/30/97	Page 2 of 3	

4.8   At the conclusion of the audit, the auditor fills in the report section of the Quality Assurance Audit Plan and Report (Figure B-3) as follows: (8) the name of the person contacted, (9) the actual date of the audit, (10) the audit findings, (11) an indication of whether or not it was necessary to issue a Corrective Action Request (see QP 3.5.1, Corrective Action Process), and (12) the signature of the auditor and the date.

4.9   The report is reviewed by Audit supervision and after any corrections, is approved by signature (13) and date.

5.0   The auditor will periodically follow up on Corrective Action Requests to verify compliance (see QP 3.5.1, Corrective Action Process).

5.1   The quality department is responsible for setting up and maintaining the appropriate files for quality audits.

SUBJECT: QUALITY ASSURANCE AUDITS OF PRODUCTS AND SYSTEMS	
Authorized By: *V. A. Philis*	**Number:** 2.2.1
Effective Date: 2/10/97 **Revision:** 1/30/97	**Page** 3 **of** 3

Figure B-3. Quality Assurance Audit Plan and Report

PLAN   Number: _____(1)

Program or product (2)	Planned audit date (3)
Audit subject/department (4)	
Detailed plan: (5)	
Plan prepared by: (6) Date:	Approved by: (7)

Report

Person contacted: (8)	Actual audit date (9)
Audit findings (10) (If necessary, use additional paper)	
Corrective action request issued against this report(11) Yes _____ No _____	
Report prepared by: (12) Date:	Approved by: (13)

SUBJECT: QUALITY ASSURANCE AUDITS OF SUBCONTRACTORS AND SUPPLIERS		
Authorized By: *V. A. Philis*	**Number:** 2.3.1	
Effective Date: 2/10/97	**Revision:** 1/30/97	**Page** 1 **of** 3

1.0 PURPOSE

This procedure sets forth t he method of periodically and randomly examining the quality performance of XYZ's subcontractors and suppliers.

2.0 APPLICATION

Applies specifically to subcontractors and suppliers. The quality audit of internal products and systems is covered by Quality Procedure 2.2.1, Quality Assurance Audits of Products and Systems.

3.0 DEFINITIONS

3.1   Quality audit: An official examination of products and systems that takes place on a periodic but random, unannounced basis, to verify the effectiveness of the company's quality program.

4.0 ASSOCIATED MATERIALS

4.1   Quality Assurance Audit Plan and Report for Subcontractor/Supplier, Form No. B-4.

4.2   Supplier Corrective Action, Quality Procedure 3.5.2

5.0 PROCEDURE

5.1   The quality management and administration department has the responsibility for planning and conducting quality audits of subcontractors and suppliers.

5.2   A subcontractor or supplier candidate for a quality audit will be designated by quality engineering on the basis of quality problems related to that candidate.

5.3   Audit supervision assigns the planning and reporting responsibilities for the candidate to a specific auditor.

5.4   The auditor prepares a plan using the Quality Assurance Audit Plan and Report for Subcontractor/Supplier form (see Figure B-4). The plan section of this form is filled out as follows: (1) the assigned number of the audit plan and report, (2) the name of the program or project, (3) the planned audit date, (4) the name and address of the subcontractor or supplier, (5) the audit subject, (6) the purchase order number and description, (7) a description of the detailed audit plan, and (8) the signature of the person who prepared the plan.

SUBJECT: QUALITY ASSURANCE AUDITS OF SUBCONTRACTORS AND SUPPLIERS		
Authorized By: *V. A. Philis*		**Number:** 2.3.1
Effective Date: 2/10/97	**Revision:** 1/30/97	**Page** __2__ **of** __3__

5.5   Audit department supervision reviews the plan and, following corrections, if any, approves the plan by (9) signing and dating the form.

5.6   The plan is now coordinated with the cognizant buyer who, following corrections, if any, approves the plan by (10) signing and dating the form.

5.7   If it is determined by quality assurance and purchasing that the subcontractor should be given advance warning of the audit, this is done by the purchasing department at this point.

5.8   The audit now takes place, including an evaluation of the production methods, workmanship, adequacy, completeness and currentness of the documentation, the effectiveness of the inspection and testing process, as well as the effectiveness of the inspectors and testers.

5.9   The auditor now prepares the report section of the Quality Assurance Audit Plan and Report for Subcontractor/ Supplier (Figure B-4) as follows: (11) the name of the subcontractor or supplier person contacted, (12) the actual audit date, (13) the audit findings, (14) an indication of whether or not it was necessary to issue a Supplier Correction Action Request (see QP 3.5.2, Supplier Corrective Action), and (15) the signature of the auditor and the date.

5.10   The report is reviewed by audit supervision and, after corrections, if any, is approved by (16) signing and dating the form.

5.11   The form is now coordinated with purchasing. Following corrections, if any, purchasing indicates approval by (17) signing and dating the form. Purchasing is responsible for transmitting the report to the subcontractor or supplier.

5.12   The auditor will periodically follow up on Supplier Corrective Action Requests to verify compliance (see QP 3.5.2, Supplier Corrective Action).

SUBJECT: QUALITY ASSURANCE AUDITS OF SUBCONTRACTORS AND SUPPLIERS		
Authorized By: *V. A. Philis*		**Number:** 2.3.1
Effective Date: 2/10/97	**Revision:** 1/30/97	**Page** 3 **of** 3

Figure B-4. Quality Assurance Audit Plan and Report for Subcontractor/Supplier

PLAN   Number: _____(1)

Program or product (2)	Planned audit date (3)
Subcontractor Supplier name and address (4)	Audit subject (5)
	Purchase order and description (6)

Detailed plan:

(7)

Report

Prepared by: (8) Date:	Approved by (QA): Date: (9)	Approved by (purchasing): Date: (10)

Report

Person contacted (11)	Actual audit date (12)

Audit findings

(13)

(If necessary, use additional paper)

Supplier Corrective Action Request issued against this report   Yes____   No____   (14)

Prepared by : (15) Date:	Approved by (QA): (16) Date:	Approved by purchasing: (17) Date:

SUBJECT: QUALITY ASSURANCE TRAINING AND CERTIFICATION PROGRAM		
Authorized By: *V. A. Philis*		**Number:** 2.4.1
Effective Date: 2/10/97	**Revision:** 1/30/97	**Page** 1 **of** 5

1.0 PURPOSE

To establish the requirements for training, certifying, and recertifying employees involved in performing critical and specialized production and inspection functions at XYZ company.

2.0 APPLICATION

Applies to employees performing certain kinds of critical or specialized functions related to deliverable items. These functions include the following (plus any additional ones that may be specified by customer contract): welding, soldering testing, nondestructive testing, and inspection. Certification is required for new employees in these categories. Recertification is required annually for all employees in these categories and more often if procedures or operating instructions change or if an employee has been decertified as a result of continued errors or problems of quality.

3.0 ASSOCIATED MATERIALS

3.1   Quality Assurance Certification/Recertification Notice, Form No. B-5

3.2   Quality Certification Card, Form No. B-6

3.3   Quality Certification Record, Form No. B-7

4.0 PROCEDURE

4.1   The responsibility for developing training courses and testing criteria, operating the training sessions, and conducting certification and recertification sessions falls under the jurisdiction of the quality management and administration department.

4.2   The quality management and administration department is also responsible for examining customer-imposed specifications related to employee certification and accommodating these requirements.

4.3   Notification to employees of quality assurance certification or recertification sessions will be accomplished through the use of the form displayed in Figure B-5. It is to be filled out as follows: (1) the employee's name, (2) the employee's identification number, (3) the department to which the employee is assigned, (4) the date the notice is prepared, (5) a check mark to indicate that this is either for certification or recertification, (6) the scheduled date, (7) time and (8) location where the process will take place (this could be at the employee's workstation), (9) the name of the person who has prepared the form, (10) that person's department, and (11) telephone number.

4.4   An employee unable to attend or be present at the session as scheduled is to have his or her supervisor arrange a new date by contacting the person who prepared the notice.

SUBJECT: QUALITY ASSURANCE TRAINING AND CERTIFICATION PROGRAM		
Authorized By: *V. A. Philis*		**Number:** 2.4.1
Effective Date: 2/10/97	**Revision:** 1/30/97	**Page** 2 **of** 5

4.5   The sessions are to be conducted by quality management and administration training and certification personnel and are to cover familiarity with the applicable specifications, procedures, and instructions, the performance of tasks related to the job classification, and the technique for maintaining records and preparing reports.

4.6   Employees who have been certified are to receive a Certification Card, the form for which is shown in Figure B-6. It is to be filled out by the Quality Assurance certifier as follows: (1) the expiration date of the certification (in no case later than one year from the issue date), (2) the function covered by this certification (i.e., "welding" or "soldering," etc.), (3) the certification number, (4) the issue date of this certification, (5) the name of the employee who has been certified, (6) the employee's identification number, (7) the department to which the employee is assigned, (8) the identification of the specification in which requirements for this function have been given (if applicable), and (9) the signature of a quality assurance official. On completion of the training, the certificate will be sent to the personnel department of the XYZ company for appropriate distribution to the employee.

4.7   The personnel department of the XYZ company is responsible for maintaining records of the certification/recertification program. The Quality Certification Record form shown in Figure B-7 is to be used for this purpose. It is to be filled out as follows: (1) the employee's name (last name first), (2) the employee's identification number, (3) the department to which the employee is assigned, (4) the employee's job classification, (5) the certifiable function being covered by this record, (6) the dates and related comments concerning this employee's training and certification sessions, (7) the dates and related comments concerning the certification expiration dates for this employee, and (8) the dates and comments related to audits made of this employee's performance. This information is going to be part of the employee's file.

4.8   Periodically, and at random, quality management and administration department training and certification personnel are to audit the performance of certified personnel and report the findings on the previously described Quality Certification Record. As a result of this audit, an employee can be discharged or required to be retrained and recertified (regardless of the expiration date of his or her current certification) if the amount of errors and/or problems with quality warrant such action.

4.9   Training will also be provided for statistical process control, failure and mode effect analysis, cost of quality, gage reproducibility and repeatability studies, capability analysis, and any other advanced statistical methodology that provides improved quality (methodologies to be considered are quality functions deployment, critical path, R&M, design of experiments, and Taguchi experimentation).

SUBJECT: QUALITY ASSURANCE TRAINING AND CERTIFICATION PROGRAM		
Authorized By: *V. A. Philis*		**Number:** 2.4.1
Effective Date: 2/10/97	**Revision:** 1/30/97	**Page** 3 **of** 5

Figure B-5. Quality Assurance Certification/Recertification Notice

To: (employee) (1)	Employee number: (2)
Department: (3)	Date: (4)

(5)

You have been scheduled for _____ certification _____ recertification
for the following function:

_____

Date: _____(6)_____   Time:_____(7)_____

Location:_____(8)_____

All certifiable functions require the certification of employees who perform these
functions. This certification must take place when the employee is hired and
thereafter on an annual basis or when there is a significant change in procedures
or instructions that relate to that function. If the above date conflicts with an
important activity, please have supervisor contact the undernamed person.

Name (9)	Department (10)	Phone (11)

SUBJECT: QUALITY ASSURANCE TRAINING AND CERTIFICATION PROGRAM		
Authorized By: *V. A. Philis*		**Number:** 2.4.1
Effective Date: 2/10/97	**Revision:** 1/30/97	**Page** 4 **of** 5

Figure B-6. Quality Certification Card Form

CERTIFICATION   Expiration date: _____ (1)

Function _____ (2)

                                                            Issue date _____ (4)

Certification number _____ (3)

This certifies that

Name _____(5)_____

Employee number ____(6)____   Department ____(7)____

Has completed requirements per _____(8)_____

_____(9)_____
Quality assurance

SUBJECT: QUALITY ASSURANCE TRAINING AND CERTIFICATION PROGRAM			
Authorized By: *V. A. Phitis*		**Number:** 2.4.1	
Effective Date: 2/10/97	**Revision:** 1/30/97	**Page** 5 **of** 5	

Figure B-7. Quality Certification Record

Quality certification record	
Name (Last, Middle, First) (1)	Department (3) Job classification (4)
Employee No. (2)	Certifiable function (5)

Training/certification dates and comments (6)

_____
_____
_____
_____
_____

Certification expiration dates and comments (7)

_____
_____
_____
_____

Certification audit dates and comments (8)

_____
_____
_____
_____

SUBJECT: QUALITY ASSURANCE FUNCTIONS		
Authorized By: *V. A. Philis*		**Number:** 3.1.1
Effective Date: 2/10/97	**Revision:** 1/30/97	**Page** __1__ **of** __2__

1.0 PURPOSE

To provide an overview of the quality assurance organization's functional involvement in the total life cycle of a product.

2.0 APPLICATION

For the general knowledge of all company personnel.

3.0 PROCEDURE

In general, the quality assurance (QA) organization is responsible for establishing acceptable limits of variations in the characteristics of a product and monitoring and reporting on compliance to these limits. This requires a number of separate, specific functions that must be taken throughout a product's life cycle (see Sections 3.1 and 3.2):

3.1   The Product Is Designed. QA function: Evaluate the evolving design to determine if the desired quality is being maintained.

3.2   Its (Manufacturing) Process Is Developed. QA function: Evaluate the proposed manufacturing process to see if the specified quality can be produced.

Product Life Cycle	QA Involvement
1) The product is designed	Evaluate the evolving design to determine if desired quality is being maintained
2) Its manufacturing process is developed	Evaluate the proposed manufacturing process to see if the specified quality can be produced
3) Its production is planned	Prepare general quality procedures and specific inspection instructions
4) Tools and equipment are designed, fabricated, procured	Inspect, calibrate, and control tools, instruments and gages when received on a continuing basis

SUBJECT: QUALITY ASSURANCE FUNCTIONS		
Authorized By: _V. A. Philis_		**Number:** 3.1.1
Effective Date: 2/10/97	**Revision:** 1/30/97	**Page** 2 **of** 2

5) Raw materials and parts are ordered	Evaluate candidate suppliers' abilities to produce to the quality that is specified
6) Materials are received	Inspect materials at their source and/or when they are received
7) The product is produced	Inspect product at various stages during its production. Analyze rejected parts and order their rework or scrap
8) The product is shipped	Perform final inspection/test of the product prior to shipment
9) The product is used	Evaluate effectiveness of product's quality. Investigate complaints.

3.3   Its Production Is Planned. QA function: Prepare general quality procedures and specific inspection instructions.

3.4   Tools and Equipment Are Designed, Fabricated, and Manufactured. QA function: Inspect, calibrate, and control tools, instruments, and gages when received on a continuing basis.

3.5   Raw Materials and Parts Are Ordered. QA function: Evaluate candidate suppliers' abilities to produce to the quality that is specified.

3.6   Materials Are Received. QA function: Inspect materials at their source and/or when they are received.

3.7   The Product Is Produced. QA function: Inspect product at various stages during its production. Analyze rejected parts and order their rework or scrap.

3.8   The Product Is Shipped. QA function: Perform final inspection/test of the product prior to shipment.

3.9   The Product Is Used. QA function: Evaluate effectiveness of product's quality. Investigate complaints (feedback data for next product cycle).

SUBJECT: INSPECTION FUNCTION LOCATIONS		
Authorized By: *V. A. Philis*		**Number:** 3.1.2
Effective Date: 2/10/97	**Revision:** 1/30/97	**Page** 1 **of** 2

1.0 PURPOSE

To provide a plant view of the XYZ company showing the physical locations of all inspection stations (see Figure B-8, Inspection Locations). The plant is located at 63 Paliocastro, Thisvi, Michigan.

2.0 APPLICATION

For the use of all company personnel who need to know the location of inspection stations.

SUBJECT: INSPECTION FUNCTION LOCATIONS		
Authorized By: *V. A. Philis*		**Number:** 3.1.2
Effective Date: 2/10/97	**Revision:** 1/30/97	**Page** _2_ **of** _2_

Figure B-8. Inspection Locations

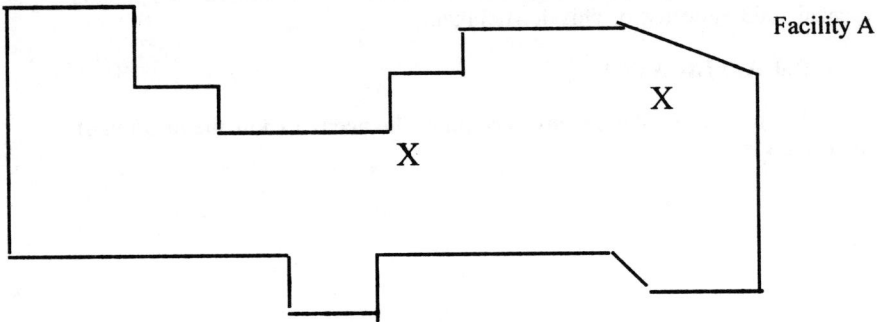

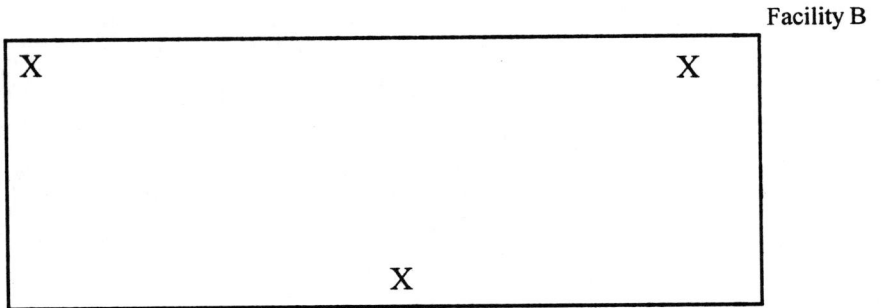

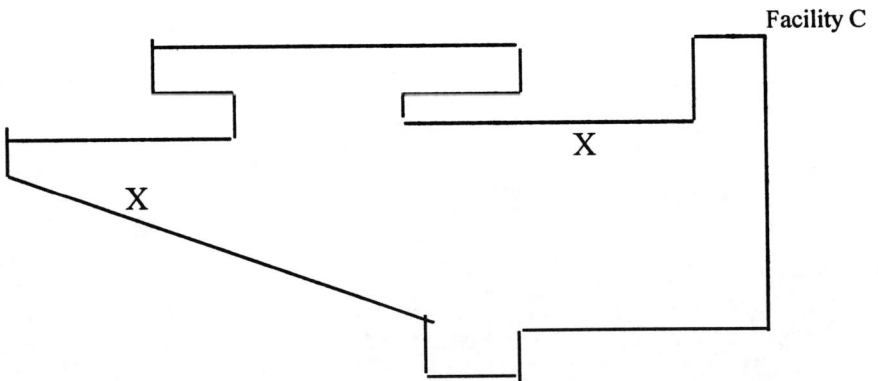

Facility A

Facility B

Facility C

SUBJECT: NEW OR PENDING CONTRACT QUALITY REQUIREMENTS ANALYSIS		
Authorized By: *V. A. Philis*	**Number:** 3.2.1	
Effective Date: 2/10/97	**Revision:** 1/30/97	**Page** 1 **of** 3

1.0 PURPOSE

To provide a means (on request from estimating or marketing department) for quickly determining the quality requirements of a new contract or a potential contract as a first step in the quality planning function at XYZ company.

2.0 APPLICATION

Relates to new and pending business and involves marketing's contracts and quality departments as required.

3.0 ASSOCIATED MATERIALS

3.1   New or Pending Contract Quality Requirements Analysis, Form No. B-9

3.2   Design Review Participation, Quality Procedure 4.1.2

4.0 PROCEDURE

4.1   The marketing contracts department of the XYZ company is responsible for providing the quality department with copies of new contracts and information regarding pending contracts.

4.2   The contracts and pending contract information are to be evaluated by the project manager from the standpoint of the impact the new business will have on quality assurance functions.

4.3   As a result of the analysis, the New or Pending Contract Quality Requirements Analysis Form shown in Figure B-9 is to be prepared. It is to be filled out as follows: (1) the name of the customer or prospect, (2) the contract number, (3) the contract type, (4) a description of the product or products being ordered, (5) the start date of the contract, (6) the end date, (7) the number of units ordered, (8) not applicable, (9) a check-mark indication if this new business is identical to, or similar to, any previous business the company has had, (10) if identical or similar, then an identification of that previous business, (11) a check-mark indication as to the contract's clarity in specifying quality requirements and then a series of statements relative to any special or unusual quality-related requirements that the new business will impose. These statements are broken down in terms of (12) a summary; (13) advanced technology techniques; (14) special inspection equipment, fixtures or gages; (15) and (16) special or unusual equipment and skills; (17) workload and personnel; (18) any special specific R&M techniques; and (19) other. The report, when complete, is (20) signed and dated by the analyst and (21) signed and dated by the analyst's supervisor following a review.

SUBJECT: NEW OR PENDING CONTRACT QUALITY REQUIREMENTS ANALYSIS		
Authorized By: *V. A. Philis*		Number: 3.2.1
Effective Date: 2/10/97	Revision: 1/30/97	Page 2 of 3

4.4   This form will be filed with the quotation and will be placed in the job folder if a purchase order is issued.

4.5   The project manager is responsible for monitoring compliance to the tasks, assignments, and dates related to acquiring the new quality resources.

4.6   If design reviews are conducted in relationship to a new or pending contract, see QP 4.1.2, Design Review Participation.

SUBJECT: NEW OR PENDING CONTRACT QUALITY REQUIREMENTS ANALYSIS			
Authorized By: *V. A. Philis*		**Number:** 3.2.1	
Effective Date: 2/10/97	**Revision:** 1/30/97	**Page** 3 **of** 3	

Figure B-9. New or Pending Contract Quality Requirements Analysis

Customer or project(1)	Contract No.(2)	Contract type (3)
Product description (4)		
Start date (5)	End date (6)	Quantity (7)	Maximum rate/mo. (8)

Critical to or similar to previous contract(s) (9) _____ Ident. _____ No _____ Similar

If identical or similar, explain (10)

Are quality terms clear (11) ____Yes ____No ____Difficult to interpret ____Unclear

Enter special or unusual quality-related requirements (use extra sheets if necessary)

Summary statement (12)

Advanced metrology techniques (13)

Special instruction equipment/fixtures/gages (14)

Special test equipment (15)

Special or unusual skills (16)

Workload/personnel (17)

Special R&M techniques (18)

Other (19)

Analysis made by (20) Date	Reviewed by (21) Date

SUBJECT: QUALITY PLANNING		
Authorized By: *V. A. Philis*		**Number:** 3.2.2
Effective Date: 2/10/97	**Revision:** 1/30/97	**Page** 1 **of** 4

1.0 PURPOSE

To establish a method at the XYZ company for developing detailed plans that cover the quality aspects of producing and delivering a product.

2.0 APPLICATION

Applies to all production programs and involves the engineering and production planning and control departments (which are responsible for providing input) and the quality assurance quality planning department.

3.0 ASSOCIATED MATERIALS

3.1   New or Pending Contract Quality Requirements Analysis (see QP 3.2.1)

3.2   Quality Bulletins (see QP 2.1.2)

3.3   Customer's standards and requirements

3.4   Test Plan (provided by the engineering department)

3.5   Quality Plan, Form Nos. B-10 and B-11

4.0 PROCEDURE

4.1   Procure, examine, and derive requirements from the New or Pending Contract Quality Requirements Analysis that relate to the job being planned, if one exists.

4.2   Receive from production planning and examine the specific manufacturing outlines that describe the production operations, the sequence of operations, and the test and inspection points related to the job that is being planned. Determine inspection methods and the need for inspection aids and gages.

4.3   Receive from engineering and examine the part drawings, specifications, and Test Plan to determine inspection requirements and detailed test requirements, test equipment and test fixtures requirements, and test equipment calibration and maintenance requirements.

4.4   Prepare Quality Plan using the forms illustrated in Figures B-10 and B-11 depending on the job requirement. The forms are to be filled out as follows: (1) the plan number, (2) the page number, (3) the program or project number, (4) the contract number, (5) the part or assembly name and (6) number, (7) the name of the

SUBJECT: QUALITY PLANNING		
Authorized By: *V. A. Philis*		**Number:** 3.2.2
Effective Date: 2/10/97	**Revision:** 1/30/97	**Page** 2 **of** 4

person preparing the plan and the date, (8) the manufacturing operation number (derived from the Manufacturing Outline), (9) the inspection/test point number (in sequence, starting with "1"), (10) the attributes to be verified including the methods to be used, (11) an identification of inspection or test equipment required for inspecting or testing each attribute, and (12) the estimated personnel hours per inspection/test point.

4.5   In preparing the Quality Plan, specify, if needed, proper handling, preservation, storage, packaging, and shipping instructions to protect the product's quality and to prevent damage, deterioration, and degradation.

4.6   Prior to release, the completed Quality Plan is to be reviewed with appropriate personnel in engineering and in production planning and control to verify the plan's compatibility with the manufacturing and test plans. When verified, it is to be approved by quality assurance management; this approval is indicated by the signature and date in the appropriate space (13) on the Quality Plan.

4.7   After approval, the Quality Plan is released for use in the preparation of detailed Inspection/Test Instructions (see Quality Procedure 3.3.1).

4.8   The quality department is responsible for maintaining a comprehensive cross-referenced file of all Quality Plans.

SUBJECT: QUALITY PLANNING		
Authorized By: *V. A. Philis*		**Number:** 3.2.2
Effective Date: 2/10/97	**Revision:** 1/30/97	**Page** _3_ **of** _4_

Figure B-10. Quality Plan

Plan number (1)		Page _____ of _____ (2)		
Program or project (3)		Contract number (4)		
Part or tool name (5)		Number (6)		
Planned by (7)		Approved by (13)		Date
Manufacturing operator number	Inspection/ test point number	Attributes to be verified and method	Inspection test equipment	Estimated hours per point
(8)	(9)	(10)	(11)	(12)

SUBJECT: QUALITY PLANNING	
Authorized By: *V. A. Philis*	**Number:** 3.2.2
Effective Date: 2/10/97 **Revision:** 1/30/97	**Page** 4 **of** 4

Figure B-11. Quality Plan Continuation Sheet

Plan number (1)			Page _____ of _____ (2)	
Manufacturing operator number	Inspection/ test point number	Attributes to be verified and method	Inspection test equipment	Estimated hours per point
(8)	(9)	(10)	(11)	(12)

SUBJECT: INSPECTION/TEST INSTRUCTIONS		
Authorized By: *V. A. Philis*		**Number:** 3.3.1
Effective Date: 2/10/97	**Revision:** 1/30/97	**Page** 1 **of** 2

1.0 PURPOSE

To provide a system for preparing, issuing, and maintaining job instructions for the inspection and testing of specific parts, subassemblies, and assemblies at the XYZ company.

2.0 APPLICATION

Applicable to personnel of the quality department of the XYZ company who are responsible for the preparation of detailed job instructions relative to monitoring product quality factors.

3.0 ASSOCIATED MATERIALS

3.1   Quality Plans, described in Quality Procedure 3.2.2

3.2   Test Procedures (provided by the engineering department)

3.3   Machined Parts Inspection Instructions, Form B-12

4.0 PROCEDURE

4.1   Obtain and review the quality department's Quality Plans and Bulletins for matching equipment types for which inspection and test instructions are to be prepared.

4.2   Prepare if required by the Quality Plan or Machined Parts Inspection Instructions form (shown in Figure B-12) as follows: (1) the part number, (2) the part name, (3) the name or number of any special fixture or gage that may be needed for the inspection, (4) the sampling criteria to be used, (5) the inspection identification method, (6) the inspection or test check point, (7) the characteristics to be inspected, (8) the reference document or documents to be used, and (9) the inspection method to be used. The inspector fills out (10) the total quantities examined versus (11) those found to be defective, (12) the date or dates of inspection, (13) the inspection stamp mark for each characteristic, and (14) remarks.

4.3   In preparing instructions, specify, if needed, proper handling, preservation, storage, packaging, and shipping instructions to protect the product's quality and to prevent damage, deterioration, and degradation.

4.4   The completed instruction is to be reviewed by quality assurance management and (15) signed and (16) dated, then issued for application.

4.5   The quality department has the responsiblity for maintaining a comprehensive cross-referenced file of all inspection and test instructions.

SUBJECT: INSPECTION/TEST INSTRUCTIONS			
Authorized By: *V. A. Philis*		**Number:** 3.3.1	
Effective Date: 2/10/97	**Revision:** 1/30/97	**Page** 2 **of** 2	

Figure B-12. Machined Parts Inspection Instructions

Part No. (1)   Part name (2)   Special fixtures or gages (3)

Sampling criteria - check one: (4) _____ Critical, 100% insp. _____ Major, insp. ___ % _____ Minor, insp. ____%

Inspection I.D. (5)

Check point	Characteristic	Reference material	Inspection method	Quantities		Inspection date	Inspection stamp
				Examined	Defective		
(6)	(7)	(8)	(9)	(10)	(11)	(12)	(13)

Remarks (14)

SUBJECT: QUALITY ASSURANCE RECORDS		
Authorized By: *V. A. Phitis*		**Number:** 3.4.1
Effective Date: 2/10/97	**Revision:** 1/30/97	**Page** 1 **of** 1

1.0 PURPOSE

To describe a system for maintaining and using complete and reliable quality program records and data at the XYZ company.

2.0 APPLICATION

Applies to all quality program records and data.

3.0 ASSOCIATED MATERIALS

3.1   Quality Assurance Reports, Quality Procedure 3.4.2

4.0 PROCEDURE

4.1   The responsibility for the operation of the quality assurance records systems is that of the documentation, data analysis, and reporting (DDAR) unit of the quality management and administration department.

4.2   The data in this system include, but are not limited to, Receiving Inspection Reports, Final Inspection and Test Reports, Failure Reports, Quality Assurance Audit Reports, and Corrective Action Requests/Reports.

4.3   The quality information is to be accumulated, organized, analyzed, and reported (see QP 3.4.2, Quality Assurance Reports).

4.4   DDAR is to establish and maintain active files for this information, together with a cross-referenced indexing system that will enable convenient search and retrieval of specific data.

4.5   Unless otherwise specified, all data are to be maintained in these active files for a period of 12 months.

SUBJECT: QUALITY ASSURANCE REPORTS		
Authorized By: *V. A. Philis*		**Number:** 3.4.2
Effective Date: 2/10/97	**Revision:** 1/30/97	**Page** 1 **of** 4

1.0 PURPOSE

To describe the method and responsibilities for compiling and organizing quality assurance data, analyzing it and generating reports that accurately describe status to bring about appropriate management actions when required at the XYZ company.

2.0 APPLICATION

Applies to both internal and contractually required reports concerning product quality. Presently, at the request of the quality manager or the affected department manager, the following categories are tracked by Pareto charts. There are plans for segregating the defects into reports.

3.0 ASSOCIATED MATERIALS

3.1   Critical Quality Problem, Form No. B-13

3.2   Defect Description, Form No. B-14

4.0 PROCEDURE

4.1   The documentation, data analysis, and reporting (DDAR) unit of the quality management and administration department has the prime responsibility for compiling, organizing, analyzing, and reporting quality assurance information.

4.2   DDAR will determine what quality information any individual or department must have in order to be able to satisfy quality requirements, whether these requirements are expressed or implied.

4.3   DDAR will determine which individuals and departments are recording quality information.

4.4   DDAR will equate the supply and demand of quality information, collecting data where recorded, specifying new data-generation requirements when needed, transforming recorded data into finished data, and disseminating data to areas required. This data may be collected using Form B-13.

4.5 In collecting the data on a Defect Description form Figure B-14, the inspector prepares it in the following manner: (1) department, (2) inspector, (3) period, (4) process charted, (5) page, (6) the defect description, and (7) the number of defects found.

SUBJECT: QUALITY ASSURANCE REPORTS		
Authorized By: *V. A. Philis*		**Number:** 3.4.2
Effective Date: 2/10/97	**Revision:** 1/30/97	**Page** 2 **of** 4

4.6   In analyzing data, when critical abnormalities requiring special actions are detected, DDAR is required to prepare and distribute a Critical Quality Problem Report, the form for which is shown in Figure B-13. This is to be filled out as follows: (1) a description of the critical problem, (2) the actual or potential impact of the problem, (3) the action that is required, (4) the name of a person who is assigned the responsibility for taking the specified action, (5) that person's department, (6) the date when the action is to be completed, (7) the dated signature of the person making the assignment, and (8) the dated signature of the person approving the assignment. Space (9) is for a later follow-up report by the person who made the assignment.

4.7   DDAR and the project managers are responsible for preparing and delivering contractually required quality reports, the terms and conditions for these dependent on the contracts involved.

4.8   The quality department is responsible for preparing and disseminating a monthly management report on quality. The report will include, but not be limited to, a Pareto chart and/or a Critical Quality Problem and/or Corrective Action Log. In each application, the forms and charts are to be subtitled to identify the specfic application (part, assembly, program, project, and/or customer) to which they apply.

4.9   The inspector, leader, and/or the project manager has the authority to initiate a Critical Quality Problem Form.

SUBJECT: QUALITY ASSURANCE REPORTS		
Authorized By: *V. A. Philis*		**Number:** 3.4.2
Effective Date: 2/10/97	**Revision:** 1/30/97	**Page** 3 **of** 4

Figure B-13. Critical Quality Problem
Requires action now

Critical problem (1)		
Actual or potential impact (2)		
Action required (3)		
Responsible for action (4)	Department (5)	Get well date (6)
Assigned by (7) Date	Approved by (8) Date	
Follow-up (9)		

SUBJECT: QUALITY ASSURANCE REPORTS		
Authorized By: *V. A. Philis*		**Number:** 3.4.2
Effective Date: 2/10/97	**Revision:** 1/30/97	**Page** __4__ **of** __4__

Figure B-14. Defect Description

Department (1)	Inspection (2)
Period (3)	Page ____ of _____ (5)
Process charted (4)	

Defect description	Number of defects found
(6)	(7)

SUBJECT: CORRECTIVE ACTION PROCESS		
Authorized By: *V. A. Philis*		Number: 3.5.1
Effective Date: 2/10/97	Revision: 1/30/97	Page 1 of 4

1.0 PURPOSE

To specify a system for informing appropriate personnel at the XYZ company of instances of nonconformance to quality requirements and to initiate corrective actions.

2.0 APPLICATION

This system encompasses the activities of all departments in the XYZ company. Its operation is the responsibility of the quality department. Corrective actions related to suppliers are covered in Quality Procedure 3.5.2, Supplier Corrective Action.

3.0 ASSOCIATED MATERIALS

3.1   Corrective Action Request, Form No. B-15

3.2   Corrective Action Status Log, Form No. B-16

4.0 PROCEDURE

4.1   A Corrective Action Request (CAR) can be initiated to cause an investigation of the cause of a material discrepancy resulting in the recommendation of corrective action to avoid the recurrence of the discrepancy if the problem is of concern as defined by management.

4.2   The generation of a CAR can come about as a result of material review board activity (see Quality Procedure 6.5.3, Material Review Process) or on receipt of either an Inspection Report or Failure Report (see Quality Procedure 6.3.1, First Article Inspection—In-House Material, and Quality Procedure 6.2.4, Failure Reporting).

4.3   The Corrective Action Request (CAR) Form (see Figure B-15) is to be filled out as follows: (1) the number of the CAR; (2) the name, address, and telephone number of the organization responsible for investigating the cause and taking corrective action; (3) the name of the person and organization requesting the action; (4) the date of the request; (5) the date when a reply to the request is due; (6) the name of the program or project; (7) the part name; (8) the part number; (9) the inspection or failure report number; (10) the Material Review Report number, if applicable; (11) a description of the condition; (12) the apparent cause of the condition, if known; (13) a description of the actual cause; (14) actions taken to remove the cause, both short term and long term; and (15) signature, title, and date.

SUBJECT: CORRECTIVE ACTION PROCESS		
Authorized By: *V. A. Philis*		**Number:** 3.5.1
Effective Date: 2/10/97	**Revision:** 1/30/97	**Page** 2 **of** 4

4.4   Information relative to the CAR is now recorded on the Corrective Action Status Log Form (see Figure B-16) as follows: (1) the CAR number, (2) the Inspection or Failure Report number, (3) the Material Review Report number if applicable, (4) the name of the organization responsible for the investigative and corrective action, (5) the date the assignment was made to that organization, and (6) the date when that organization's response is due.

4.5   The CAR is forwarded to the assigned organization, where an investigation is done to detect the cause of the discrepancy and action is taken to prevent recurrence. This information is recorded on CAR form as follows: (13) actual cause of the discrepancy, (14) action taken to prevent recurrence, and (15) the signature and title of the person responsible for the corrective action and the date of the signature.

4.6   The completed CAR is returned to the quality department. The Corrective Action Status Log is updated to reflect (7) the date of the actual response and (8) remarks pertinent to the action.

SUBJECT: CORRECTIVE ACTION PROCESS		
Authorized By: *V. A. Philis*		**Number:** 3.5.1
Effective Date: 2/10/97	**Revision:** 1/30/97	**Page** _3_ **of** _4_

Figure B-15. Corrective Action Request

Number (1)	To (organization name, address, phone number) (2)	
From (3)	Request date (4)	Reply due date (5)
Program/project (6)	Part name (7)	Part number (8)
Inspection/failure report number (9)	MRR Number (10)	
Description of condition (11)		
Apparent cause (12)		
Actual cause (13)		
Action taken to prevent recurrence (14)		
Signature (15) Title Date		

SUBJECT: CORRECTIVE ACTION PROCESS							
Authorized By: *V. A. Philis*					**Number:** 3.5.1		
Effective Date: 2/10/97		**Revision:** 1/30/97			**Page** _4_ **of** _4_		

Figure B-16. Corrective Action Status Log

CAR NO	Inspection or Failure Report No.	Material Review Report No.	Organization responsible for action	Date assigned	Date response due	Date of actual response	Remarks
(1)	(2)	(3)	(4)	(5)	(6)	(7)	(8)

SUBJECT: SUPPLIER CORRECTIVE ACTION		
Authorized By: *V. A. Philis*	**Number:** 3.5.2	
Effective Date: 2/10/97	**Revision:** 1/30/97	**Page** 1 **of** 3

1.0 PURPOSE

To specify a system for informing suppliers of instances of nonconformance to the XYZ company's quality requirements and to initiate corrective action.

2.0 APPLICATION

This procedure relates to corrective actions related to suppliers. The system's operation is the responsibility of the quality department at the XYZ company.

3.0 ASSOCIATED MATERIALS

3.1   Supplier Corrective Action Request, Form No. B-17

4.0 PROCEDURE

4.1   When supplier corrective action is required, a Supplier Corrective Action Request (SCAR) Form (see Figure B-17) is to be filled out either by the material review board or the quality department.

4.2   The SCAR form is to be filled out as follows: (1) the SCAR number, (2) the date the form is being prepared, (3) the name and address of the supplier, (4) the name and address of the sender, (5) the part number in question, (6) the part name, (7) the quantity of parts received, (8) the quantity rejected, (9) the name of the program or project, (10) the purchase order number, (11) a description of the discrepancy, (12) a check mark as to the disposition taken, (13) the approval signature of the appropriate purchasing representative, and (14) the approval signature of the appropriate quality assurance representative.

4.3   The SCAR, whether generated by quality engineering or the material review board, is to be processed through the quality engineering unit for logging and control purposes.

4.4   After logging, the SCAR is sent to material coordination for forwarding to the supplier.

4.5   The supplier investigates the discrepancy and takes corrective action; this is recorded on the SCAR form as follows: (15) a description of the cause of the discrepancy, (16) a description of the corrective action taken to eliminate a recurrence of the discrepancy, and (17) the supplier's signature, title, and date.

SUBJECT: SUPPLIER CORRECTIVE ACTION			
Authorized By: *V. A. Philis*		**Number:** 3.5.2	
Effective Date: 2/10/97	**Revision:** 1/30/97	**Page** 2 **of** 3	

4.6   The completed SCAR is to be returned by the supplier, usually within a specified date. If the form has not been received after this period, purchasing is to send a reminder to the supplier. If there is no reply, then the supplier is taken off the quote list.

4.7   After receipt of the SCAR from the supplier, it is reviewed by quality engineering, which can request additional data or changed data from the supplier. Acceptable SCARs are then forwarded to the material review board.

SUBJECT: SUPPLIER CORRECTIVE ACTION		
Authorized By: *V. A. Philis*		**Number:** 3.5.2
Effective Date: 2/10/97	**Revision:** 1/30/97	**Page** 3 **of** 3

Figure B-17. Supplier Corrective Action Request

Number (1)	To (3)	From (4)
Date (2)		
Part No. (5)		
Part name (6)	Quantity received (7)	Quantity rejected (8)
Program/project (9)		Purchase order No. (10)

Description of discrepancy (11)

Disposition (12)

    ___ returned to you for evaluation   ___ returned to you for rework
    ___ used as is   ___ reworked at your expense
    ___ other (specify)

_____   _____
Purchasing approval (13)   Quality assurance approval (14)

Supplier to complete the following

Cause of discrepancy (15)

Corrective action taken to eliminate recurrence of discrepancy (16)

Supplier authorized signature   (17)   Title   Date

SUBJECT: STANDARD REPAIRS		
Authorized By: *V. A. Philis*		**Number:** 3.5.3
Effective Date: 2/10/97	**Revision:** 1/30/97	**Page** 1 **of** 1

1.0 PURPOSE

To provide a practical approach for readily repairing certain types of minor discrepancies without the need for routing the nonconforming material through the normal Material Review Process.

2.0 APPLICATION

Applies only to those items and repairs covered by normal procedures of this manual.

3.0 PROCEDURES

3.1   Nonconforming materials are identified with a Rejection Tag (see Quality Procedure 6.5.3, Material Review Process).

3.2   Tagged materials are examined in terms of the Inspection Report; they are repaired or scrapped.

3.3   Deliver nonconforming items to appropriate locations for repair.

SUBJECT: COSTS RELATED TO QUALITY		
Authorized By: *V. A. Philis*		**Number:** 3.6.1
Effective Date: 2/10/97	**Revision:** 1/30/97	**Page** 1 **of** 1

1.0 PURPOSE

To establish the method at the XYZ company for collecting, maintaining, and using quality cost data.

2.0 APPLICATION

Pertains to all quality costs connected with preventing and correcting nonconforming materials.

3.0 PROCEDURE

3.1   The quality management and administration department of quality assurance has the responsibility for collecting quality cost data, organizing, evaluating, and maintaining records of this information and generating quality cost reports.

3.2   The quality cost material is to be organized and summarized in four categories: (1) prevention, (2) evaluation, (3) corrective actions, and (4) customer quality problems.

3.3   Hourly and salary quality assurance personnel are to record their time charges by these four categories.

3.4   The quality costs relative to the <u>prevention</u> category are those associated with quality engineering, quality planning, quality training and indoctrination, employee certification, and candidate supplier/vendor/subcontractor evaluations.

3.5   The quality costs relative to the <u>evaluation</u> (appraisal) category are those associated with inspection, testing, quality audits, calibration and repair of test equipment, and process control.

3.6   The quality costs relative to the <u>corrective actions</u> (internal failure) category are those associated with rework, repair, scrap, reinspection, retesting, and material review.

3.7   The quality costs relative to the <u>customer quality problems</u> (external failure) category are those associated with processing complaints, repair of returned materials, replacement of returned materials, and handling and shipping.

3.8   Quality cost data are to be collected from the hourly and salary tab runs and from the scrap and rework tab runs.

3.9   Quality cost reports are to be prepared monthly, with costs reported by category, as well as summarized and compared currently and historically to total costs of goods. Distribution of the report will be controlled by the director of quality assurance.

SUBJECT: ENGINEERING DOCUMENTATION REVIEW AND CONTROL		
Authorized By: *V. A. Philis*		**Number:** 4.1.1
Effective Date: 2/10/97	**Revision:** 1/30/97	**Page** 1 **of** 3

1.0 PURPOSE

The purpose of this procedure at the XYZ company is to describe the system that quality assurance uses to review and critique engineering and supplemental documentation for adequacy, completeness, and currentness and to monitor in-house and subcontractor operations to see that approved changes are incorporated in the designated effectivity point.

2.0 APPLICATION

Applies to all released materials, including engineering drawings, specifications, process instructions, production engineering instructions, industrial engineering instructions, design-related work instructions, and engineering change orders.

3.0 ASSOCIATED MATERIALS

3.1   Design Review Participation, Quality Procedure 4.1.2

3.2   Quality Planning, Quality Procedure 3.2.2

3.3   Inspection/Test Instructions, Quality Procedure 3.3.1

3.4   Engineering Change Form, Figure B-18

4.0 PROCEDURE

4.1   The quality engineering department has the prime responsibility for engineering documentation review and control. In performing this work, quality engineering will participate in design reviews (see Quality Procedure 4.1.2) and will be the quality assurance representative on the change control board.

4.2   Engineering drawings are to be reviewed by the project manager for capability, completeness, and currentness, with corrective action taken regarding discrepancies.

4.3   Audits are to be conducted periodically and on a random basis to see that only current material is being distributed and used and that obsolete material has been removed.

4.4   Supplemental specifications such as process instructions, production engineering instructions, and design-related work instructions are to be reviewed by quality engineering for adequacy, completeness, and currentness, with corrective action taken regarding discrepancies.

SUBJECT: ENGINEERING DOCUMENTATION REVIEW AND CONTROL		
Authorized By: *V. A. Philis*		Number: 4.1.1
Effective Date: 2/10/97	Revision: 1/30/97	Page 2 of 3

4.5   Quality engineering is responsible for reviewing and analyzing inspection records and reports. One purpose is to determine quality-related design improvements in the product. These recommendations are forwarded to engineering.

4.6   As a participant on the change control board, the project manager is to review Engineering Change Requests (ECRs) and Engineering Change Orders (ECOs) to determine if they are correct, complete, and in compliance with standards and needs.

4.7   As a participant in the change control board, the project manager is to establish effectivity points for changes that have been approved.

4.8   The tool person is sometimes the individual who initiates changes in the blueprints because of difficulty in assembling the fixtures. The process is followed through the system by using Form B-18 as follows: (1) job number, (2) design vendor, (3) design contact, (4) follow-up (customer contact), (5) customer, (6) tool number, (7) tool description, (8) detail number, (9) sheet number of affected detail, (10) a brief description of additional work performed and the cause discovered, (11) date of problem discovered, (12) location of part, (13) date design was notified of the requested change, (14) date the changed drawings were received, (15) change letter, and (16) remarks.

4.9   Quality department and/or the project manager is responsible to alter Quality Plans and Inspection/Test Instructions if such alterations are required by the approved change (see Quality Procedures 3.2.2 and 3.3.1)

4.10   The project manager will monitor the incorporation of production and assembly changes at the designated effectivity points, taking corrective action when discrepancies are found.

4.11   The project manager is responsible for tabulating cost for engineering changes and categorizing such cost as to the reason for change, and supplying it to the quality department for use in the Cost of Quality Report.

SUBJECT: ENGINEERING DOCUMENTATION REVIEW AND CONTROL		
Authorized By: *V. A. Philis*		**Number:** 4.1.1
Effective Date: 2/10/97	**Revision:** 1/30/97	**Page** 3 **of** 3

Figure B-18. Engineering Change

Job No. (1)	Design supplier (2)	Design contact (3)		Follow-up person (4)				
Customer (5)	File/project No. (6)			Tool description (7)				
Detail No.	Sheet No.	Description change	Date found	Location of part	Date design told	Date drawings received	Change letter	Remarks
(8)	(9)	(10)	(11)	(12)	(13)	(14)	(15)	(16)

SUBJECT: DESIGN REVIEW PARTICIPATION		
Authorized By: *V. A. Philis*		**Number:** 4.1.2
Effective Date: 2/10/97	**Revision:** 1/30/97	**Page** 1 **of** 5

1.0 PURPOSE

Describes the quality assurance organization's participation in the design review process. The purpose of this participation is to analyze and critique new design and redesign concepts from the standpoint of quality at the XYZ company.

2.0 APPLICATION

Applies to the review of all new design and redesign concepts related to deliverable products at the XYZ company.

3.0 ASSOCIATED MATERIALS

3.1   Design Factors—End Use Quality, Form No. B-19

3.2   Design Factors—Quality-Related Producibility/Simplification Form No. B-20

3.3   Design Review Report—Product Quality Considerations, Form No. B-21

3.4   New or Pending Contract Quality Requirements Analysis, Quality Procedure 3.2.1

4.0 PROCEDURE

4.1   The project manager has the responsibility for serving as the quality assurance organization's representative at design reviews.

4.2   When the design review work is in connection with a new or pending contract, there might also be the requirement for a New or Pending Contract Quality Requirements Analysis (see Quality Procedure 3.2.1).

4.3   Quality engineering's participation in design reviews requires an analysis of quality consideration using a set of Design Factors forms. The first of these, Design Factors—End Use Quality, is shown in Figure B-19 and should be filled out as follows: (1) the name of the item for design review, (2) the item's number, (3) the review (if this is one of a series of reviews for the same program or project), (4) the date of the review, and (5) a list of quality considerations. For each consideration, there should be notations as to (6) the current design, (7) options (alternatives), and (8) action required.

SUBJECT: DESIGN REVIEW PARTICIPATION		
Authorized By: *V. A. Philis*		**Number:** 4.1.2
Effective Date: 2/10/97	**Revision:** 1/30/97	**Page** 2 **of** 5

4.4   Is there a delicate nature to the product that requires special handling during production and assembly? Will the product require unique, special packaging and shipping?   If so, a checklist is prepared as needed by the project manager for appropriate follow-up.

4.5   The next form to be used in design reviews is shown in Figure B-20, Design Factors—Quality-Related Producibility/Simplification. It should be filled out as follows: (1) item name, (2) item number, (3) review number, (4) date of the review, and (5) a list of quality considerations. Information as to (6) current design, (7) options, and (8) action required should be recorded for each quality consideration.

4.6   The project manager is required to prepare and distribute a report following each design review. The form for this report is shown in Figure B-21, Design Review Report—Product Quality Considerations. It is to be filled out as follows: (1) the name of the equipment with the design being reviewed, (2) the item number, (3) the review number, (4) the date of the review, (5) the location where the review was conducted, (6) the attendees' names and (7) organizations, (8) the product quality topics discussed, (9) a check mark to indicate whether or not a set of design factors worksheets are attached to this report, (10) actions required, (11) the names of persons assigned to take these actions, (12) the dates when the actions are to be completed, (13) the signature of the person who has prepared the report, and (14) the signature of the person who has approved the report.

SUBJECT: DESIGN REVIEW PARTICIPATION		
Authorized By: *V. A. Philis*	**Number:** 4.1.2	
Effective Date: 2/10/97 **Revision:** 1/30/97	**Page** _3_ **of** _5_	

Figure B-19. Design Factors—End Use Quality

Item name (1)	Item No. (2)	Review No. (3)	Date (4)	
Considerations (5)		Current design (6)	Options (7)	Actions required (8)
What is the product's intended quality level?				
What is the product's designed life?				
Can the product be of lesser quality and still be competitive?				
What level of maintenance is required to retain the product's quality?				
What factors might cause accelerated degradation of quality?				
Does the product, because of design, pose special hazards to users?				
Other (specify).				

SUBJECT: DESIGN REVIEW PARTICIPATION		
Authorized By: *V. A. Philis*		**Number:** 4.1.2
Effective Date: 2/10/97	**Revision:** 1/30/97	**Page** 4 **of** 5

Figure B-20. Design factors—Quality-Related Producibility/Simplification

Item name (1)	Item No. (2)	Review No. (3)	Date (4)	
Considerations (5)		Current design (6)	Options (7)	Actions required (8)
Does the design involve leading edge technology, and is it needed?				
What kinds of potential quality problems were disclosed by mock-ups, models, prototypes, and/or breadboards?				
Have tolerances been liberalized to the greatest possible extent?				
Can redesign substitute an adequate standard component for a nonstandard one?				
Are there viable alternatives to components with anticipated high failure rates?				
Would the system, as designed, be subject to chain-reaction-type failures?				
Other (specify).				

SUBJECT: DESIGN REVIEW PARTICIPATION		
Authorized By: *V. A. Philis*		**Number:** 4.1.2
Effective Date: 2/10/97	**Revision:** 1/30/97	**Page** __5__ **of** __5__

Figure B-21. Design Review Report—Product Quality Considerations

Item name (1)	Item No. (2)	Review No. (3)	Date (4)
Location (5)			
Attendees' names (6)		Organizations (7)	
Product quality topics discussed (8)			
Detailed design factors worksheets attached? (9) _____ Yes _____ No			

Actions required (10)	Assigned to (11)	Complete by (date) (12)

Prepared by (13)	Approved by (14)

SUBJECT: DELIVERABLE DOCUMENTATION REVIEW		
Authorized By: *V. A. Philis*		**Number:** 4.1.3
Effective Date: 2/10/97	**Revision:** 1/30/97	**Page** __1__ **of** __2__

1.0 PURPOSE

To describe a system for verifying the completeness and correctness of deliverable documentation.

2.0 APPLICATION

Applies to all programs and projects that require delivery of items of supportive, certifying, and explanatory documentation in accompaniment with the equipment.

3.0 ASSOCIATED MATERIALS

3.1   Deliverable Documentation Checklist, Form No. B-22.

3.2   Final Acceptance Processing, Quality Procedure 6.3.4

4.0 PROCEDURE

4.1   The responsiblity for verifying the completeness and correctness of deliverable documentation is that of the final inspection department.

4.2   This work is to be done in conjunction with final acceptance processing as described in Quality Procedure 6.3.4.

4.3   Applicable contracts, purchase orders, and other agreements are to be examined to determine for each deliverable item the documentation that is to accompany it. This information is to be recorded on the Deliverable Documentation Checklist Form, an example of which is shown in Figure B-22.

4.4   The Deliverable Documentation Checklist Form is to be filled out as follows: (1) the part name, (2) the part number, (3) the serial number, (4) the customer, (5) the purchase order or contract number, (6) the date, (7) bill of material, (8) the documentation item with check marks to indicate (9) which items are required and which items are available and ready to send, (11) remarks, (12) the dated signature of the person preparing the report, and (13) the dated signature of the person approving the report.

4.5   Individual items of documentation are to be examined to see if they are complete and, when applicable, apply to specifications for that type of documentation.

SUBJECT: DELIVERABLE DOCUMENTATION REVIEW		
Authorized By: *V. A. Philis*		**Number:** 4.1.3
Effective Date: 2/10/97	**Revision:** 1/30/97	**Page** 2 **of** 2

Figure B-22. Deliverable Documentation Checklist

Part name (1)	Part No. (2)	Serial No. (3)	
Customer (4)	P.O. or contract (5)	Date (6)	

Item No. (7)	Documentation item (8)	Check mark	
		Required (9)	Available (10)
1	Serialized assembly parts list		
2	As-built drawings		
3	Spares list		
4	Operating instructions		
5	Maintenance instructions		
6	Final inspection report		
7	Unit test report		
8	System test report		
9	Hardness tests		
10	Stress tests		
11	Weight record		
12	Capacity affidavit		
13	Certifications		
14	Nonconformance reports		
15	Other (specify)		
.			
.			
.			
.			
16	Other		

Remarks (11)	
Prepared by (12) Date	Approved by (13) Date

SUBJECT: CALIBRATION SYSTEM RESPONSIBILITIES		
Authorized By: *V. A. Philis*		**Number:** 4.2.1
Effective Date: 2/10/97	**Revision:** 1/30/97	**Page** 1 **of** 2

1.0 PURPOSE

To specify the responsibilities for controlling the accuracy of measuring the test equipment and measurement standards of the XYZ company to ensure that products delivered to customers conform to specified requirements.

2.0 PROCEDURE

2.1   All measuring and test equipment instruments and devices used to determine an item's conformance to specified requirements are to be calibrated. This calibration is to occur at regularly scheduled intervals determined on the basis of stability, purpose, and usage, or sooner if there is some reason to believe that the instrument or device needs recalibration.

2.2   All measuring and test equipment devices will be calibrated to working measurement standards or transfer measurement standards, with these in turn calibrated to reference measurement standards calibrated and certified by the National Bureau of Standards.

2.3   Records are to be maintained that identify each item of measuring and test equipment, each measurement standard, and a list and date of each instance of calibration, citing measurements and adjustments. The records are to be able to demonstrate traceability of the calibration work to the National Bureau of Standards.

2.4   Each item of measuring and test equipment and measurement standard is to be marked showing the date of the most recent calibration, the stamp of the technician who performed the calibration, and the date when the next calibration is scheduled. If the item is too small for this type of marking, the serial number is to be used for maintaining records that cite the same data.

2.5   Departments using measuring and test equipment are responsible for monitoring calibration due dates and submitting instruments and devices for calibration on schedule.

2.6   Inspectors and test technicians cannot accept measurement values obtained on measuring and test equipment that have exceeded calibration due dates.

2.7   Measuring and test equipment instruments and devices cannot be calibrated with measurement standards that have exceeded calibration due dates.

2.8   Employee-owned measuring equipment (such as micrometers) can be used for product acceptance measurements only if they are calibrated by the company on a regularly scheduled basis.

2.9   The environment in which measuring and test equipment instruments and devices are to be calibrated and used must be controlled to the extent

SUBJECT: CALIBRATION SYSTEM RESPONSIBILITIES		
Authorized By: *V. A. Philis*		**Number:** 4.2.1
Effective Date: 2/10/97	**Revision:** 1/30/97	**Page** 2 **of** 2

necessary to ensure required accuracy, with consideration given to temperature, humidity, vibration, cleanliness, and other controllable factors.

2.10   Purchasing is responsible for coordinating with quality assurance in the selection and acquisition of commercial measuring and test equipment and in the evaluation and selection of subcontractors to perform measuring and test equipment calibration and/or repair work.

2.11   All new, reworked, repaired, or modified measuring and test equipment instruments and devices are to be examined and, when proven acceptable, certified as complying to requirements by the quality department.

2.13   The purchasing department is responsible for securing calibration job instructions for all commercial measuring and test equipment that has been procured and for providing these to the quality department.

2.14   The quality department will be responsible for procuring and maintaining all measurement standards required to support product measurement requirements.

SUBJECT: CALIBRATION CONTROL AND RECALL		
Authorized By: *V. A. Philis*		**Number:** 4.2.2
Effective Date: 2/10/97	**Revision:** 1/30/97	**Page** 1 **of** 4

1.0 PURPOSE

This procedure describes the method and forms needed for calibrating and maintaining calibration records of all company measuring and test equipment devices and measurement standards.

2.0 APPLICATION

Applies specifically to the quality assurance organization's measuring equipment control (MEC) department and generally to all employees and departments who utilize measuring and test equipment.

3.0 DEFINITIONS

3.1   Calibration control: A documented system for ensuring that measuring and test equipment devices and measurement standards are calibrated, and that this is done at intervals that ensure their accuracy.

3.2   Calibration recall: A system for indicating in advance for each measuring and test equipment device and for each measurement standard the date when it is next due to be calibrated.

4.0 ASSOCIATED MATERIALS

4.1   Calibration System Responsibilities, Quality Procedure 4.2.1

4.2   Equipment Location and Recall Record, Form No. B-23

4.3   Calibration/Service Data Record, Form No. B-24

5.0 PROCEDURE

5.1   The quality department is responsible for the identification, calibration, repair, and calibration record keeping of all measuring standards. The records must offer traceability to the National Bureau of Standards.

5.2   All personnel and departments using measuring and test equipment have the responsibility for seeing that an item of equipment is not used when its calibration period has expired. Such items are to be returned to the MEC department for calibration, or arrangements must be made to secure on-site calibration of the equipment when appropriate.

SUBJECT: CALIBRATION CONTROL AND RECALL		
Authorized By: *V. A. Philis*	**Number:** 4.2.2	
Effective Date: 2/10/97	**Revision:** 1/30/97	**Page** 2 **of** 4

5.3   An Equipment Location and Recall Record Form, such as shown in Figure B-23, is to be created and maintained for each item of measurement standards, measuring equipment, and test equipment. The form is to be filled out as follows: (1) the item's name and (2) serial number, (3) the item's manufacturer, (4) the model, (5) the recall date or dates (positioned so that the form can be filed in a "tickler" file by date), (6) the calibration specification number, (7) the calibration job instruction number, (8) the calibration/service data record (CSDR) number, (9) the test or calibration frequency, (10) the type of test or calibration, and (11) the item's location. Each time the item is calibrated, the information is recorded as follows: (12) the actual recall date, (13) the date the calibration was actually performed, (14) the identifying stamp of the person who performed the calibration, (15) the date when the item was returned to use, and (16) any remarks related to that calibration cycle.

5.4   A Calibration/Service Data Record Form, such as shown in Figure B-24, is to be created and maintained for each item of measurement standards, measuring equipment, and test equipment. It is to be filled out as follows: (1) the item's name, (2) the item's manufacturer, (3) the model name and/or number, (4) the recall date or dates, (5) the calibration/service specification number, (6) the calibration/service data record (CSDR) number (to be assigned by the person filling out this form), (7) the calibration job instruction number, (8) the date this record is first created, (9) the signature of the person approving the basic record, and (10) a listing of the measurement standard or standards required to calibrate this item. Actual measurement data are recorded in terms of (11) the operation number, (12) the parameter, the (13) nominal, (14) measured, and (15) corrected reading or value, the (16) minimum and (17) maximum limits, and (18) remarks.

5.5   Calibration and repair status identification is accomplished through the use of a decal applied to each item of measurement standards, measuring equipment, and test equipment. This decal is identified in Figure B-27 and is explained in Section 4.2.4 in Paragraph 5.3.

5.6   The quality department has the responsibility for continually examining the calibration intervals assigned to the measurement devices and extending or shortening them as required.

5.7   Calibration of a measurement device can be requested at any time, regardless of the calibration due date for that device, following the occurrence of any event that places the device's accuracy in doubt.

5.8   The quality department is to maintain a surveillance function. Its purpose is to periodically and randomly audit compliance to the calibration control and recall system. Violations are to be reported to the vice president of administration.

SUBJECT: CALIBRATION CONTROL AND RECALL		
Authorized By: *V. A. Philis*		**Number:** 4.2.2
Effective Date: 2/10/97	**Revision:** 1/30/97	**Page** 3 **of** 4

Figure B-23. Equipment Location and Recall Record

Item name (1)	Serial No. (2)	Manufacturer (3)
Model (4)	Recall (5)	Calibration spec. No. (6)
Instruction No. (7)	Test or calibration frequency (9)	
Calibration service data (8)	Type of test or calibration (10)	
Item location (11)		

Recall date	Calibration date	Assign to	Returned to use	Remarks
(12)	(13)	(14)	(15)	(16)

SUBJECT: CALIBRATION CONTROL AND RECALL		
Authorized By: *V. A. Philis*		**Number:** 4.2.2
Effective Date: 2/10/97	**Revision:** 1/30/97	**Page** 4 **of** 4

Figure B-24. Calibration/Service Data Record

Item name (1)	Manufacturer (2)	Model (3)	Recall (4)
Spec. No. (5)	CSDR (6)	Job Instruction No. (7)	
Date (8)	Approved (9) Name	Title	Date

Measuring standards required **(10)**

Operator		Reading or value			Limits	
	Parameter	Nominal	Measured	Corrected	Minimum	Maximum
(11)	(12)	(13)	(14)	(15)	(16)	(17)

Remarks (18)

SUBJECT: CALIBRATION/SERVICE SPECIFICATIONS AND INSTRUCTIONS		
Authorized By: *V. A. Philis*		**Number:** 4.2.3
Effective Date: 2/10/97	**Revision:** 1/30/97	**Page** 1 **of** 3

1.0 PURPOSE

This procedure describes the methods and responsibilities for acquiring, preparing, and controlling specifications and instructions for calibrating measurement standards and measuring and test equipment of the XYZ company.

2.0 APPLICATION

Applies to the quality control department, which has prime responsibility for calibration specifications and instructions, and the purchasing department, which has associated responsibilities.

3.0 ASSOCIATED MATERIALS

3.1   Calibration System Responsibilities, Quality Procedure 4.2.1

3.2   Calibration/Service Specification, Form No. B-25

4.0 PROCEDURE

4.1   The quality department is responsible for creating and maintaining a file of calibration specifications and instructions that cover all of the company's measuring and test equipment devices and measurement standards.

4.2   The calibration specification and instruction filing systems must include serialized numbering schemes that provide unique numbers for each specification and instruction.

4.3   The purchasing department is responsible for securing from the manufacturers both the calibration/service specifications and the calibration instructions for commercial measuring devices at the time these devices are procured and providing these materials to the quality department.

4.4   The quality department is responsible for developing calibration specifications and instructions for measuring and test equipment devices measurement standards when such specifications and instructions do not exist or cannot be obtained.

4.5   The Calibration/Service Specification Form to be used for all calibration specifications is shown in Figure B-25. The information to be recorded on this form is as follows: (1) the item's name, (2) the item's manufacturer, (3) the model name and/or number, (4) the calibration job instruction (JI) number, (5) the calibration/service specification number (to be assigned by the person generating the form),

SUBJECT: CALIBRATION/SERVICE SPECIFICATIONS AND INSTRUCTIONS		
Authorized By: *V. A. Philis*		**Number:** 4.2.3
Effective Date: 2/10/97	**Revision:** 1/30/97	**Page** 2 **of** 3

(6) the calibration/service data record (CSDR) number, (7) the date that the specification is being prepared, and (8) the signature of the person who approved the specification. Calibration requirements are specified as follows: (9) the calibration source (the company or a commercial), (10) the maximum interval between calibrations, (11) a general description of calibration requirements and operating requirements, (12) a specification of environmental conditions, (13) the characteristic or range, (14) the tolerance, and (15) the condition or limitation. Service requirements are given in terms of (16) whether tamper-proof sealing is required, (17) preventive maintenance, and (18) operation test.

4.6   Calibration instructions should identify the item to be calibrated, list the calibration equipment required, provide a block diagram of the calibration setup, and provide a step-by-step method for performing the work.

SUBJECT: CALIBRATION/SERVICE SPECIFICATIONS AND INSTRUCTIONS		
Authorized By: *V. A. Philis*		**Number:** 4.2.3
Effective Date: 2/10/97	**Revision:** 1/30/97	**Page** 3 **of** 3

Figure B-25. Calibration/Service Specification

Item name (1)		JI No. (4)	Spec. No. (5)
Manufacturer (2)		CSDR No. (6)	Date (7)
Model (3)		Approved (8)	

Calibration requirements		
Source (9)	Maximum interval (10)	
General description and operating requirements (11)		
Environmental conditions (12)		

Characteristic/range (13)	Tolerance (14)	Condition/limitation (15)

Service requirements	
Tamper-proof sealing required (16) _____ Not required _____	
Preventive maintenance (17)	
Operational test (18)	

SUBJECT: CERTIFICATION OF MEASURING AND TEST EQUIPMENT AND TEST PROCESSES		
Authorized By: *V. A. Philis*		**Number:** 4.2.4
Effective Date: 2/10/97	**Revision:** 1/30/97	**Page** __1__ **of** __3__

1.0 PURPOSE

To define a method for endorsing the use of measuring and test equipment devices and processes and their associated specifications and instructions at the XYZ company.

2.0 APPLICATION

Applies to all of the XYZ company's measuring equipment, test equipment, and test processes used in determining if deliverable products conform to specified requirements.

3.0 DEFINITION

3.1   Certification: Approval given for the use of newly acquired or reworked or modified measuring and test equipment devices or processes following examinations that have verified that they are compatible with other device processes in the system and, when calibrated according to specifications, are accurate and capable of fulfilling intended funtions.

4.0 ASSOCIATED MATERIALS

4.1   Calibration System Responsibilities, Quality Procedure 4.2.1

4.2   Notice of Certification, Form No. B-26

4.3   Certification Decal, Form No. B-27

5.0 PROCEDURE

5.1   It is the responsibility of the quality department to examine and test all newly acquired, reworked, or modified measuring and test equipment devices or processes, together with their associated specifications, to determine if they are compatible with other devices and processes in the system and if they are accurate as specified.

5.2   For each device or process found to be acceptable, a Notice of Certification is to be prepared. This form, shown in Figure B-26, is to be filled out as follows: (1) the identification of the device or process being certified, (2) the date of the certification, (3) the date when the certification will no longer be valid, (4) the certification level (either restricted, interim, or final), (5) the specification number, (6) the calibration job instruction number, (7) the part number or numbers measured or tested by this device or process, (8) comments, (9) the dated signature of the person preparing the notice, and (10) the dated signature of the person approving the certification.

SUBJECT: CERTIFICATION OF MEASURING AND TEST EQUIPMENT AND TEST PROCESSES		
Authorized By: *V. A. Philis*		**Number:** 4.2.4
Effective Date: 2/10/97	**Revision:** 1/30/97	**Page** 2 **of** 3

5.3   For each device found to be acceptable, a Certification Decal, such as shown in Figure B-27, is to be prepared and applied. It is to be filled out with (1) the item name, (2) the item's serial number, (3) the model name or number, (4) the name of the manufacturer, (5) the device's usable ranges/parameters, (6) the date of the certification, and (7) the stamp of the individual who certified the device.

5.4   The quality department is responsible for publishing and distributing, on a monthly basis, a list of certified measuring and test equipment devices and processes, each item identified in terms of name, type, serial number, model name or number, and manufacturer or fabrication department name.

5.5   The quality department has the authority to decertify measuring and test equipment devices and processes when conditions occur that could adversely affect accuracy.

SUBJECT: CERTIFICATION OF MEASURING AND TEST EQUIPMENT AND TEST PROCESSES		
Authorized By: *V. A. Philis*		**Number:** 4.2.4
Effective Date: 2/10/97	**Revision:** 1/30/97	**Page** 3 **of** 3

Figure B-26. Notice of Certification

Certification of (1)	Date (2)
	Expiration date (3)
	Certification level (4) Restricted _____ Interim _____ Final _____
Specification No. (5)	Calibration job instruction No. (6)
Part number(s) (7)	
Comments (8)	
Prepared by (9)　　　Date　　　Approved by (10)　　　Date	

Figure B-27. Certification Decal

Item (1)	Model (3)
Serial No. (2)	Manufacturer (4)
Usable ranges/parameters (5) 　　Nominal　　　Indicated　　　　Tolerance	
Date (6)　　　　Stamp or approval signature (7)	

SUBJECT: CALIBRATION AND CONTROL OF DIMENSIONAL WORK- ING INSTRUMENTS		
Authorized By: *V. A. Philis*		**Number:** 4.2.5
Effective Date: 2/10/97	**Revision:** 1/30/97	**Page** 1 **of** 2

1.0 PURPOSE

To describe the system for periodic calibration of dimensional working instruments at the XYZ company.

2.0 APPLICATION

Applies to the calibration control of dimensional working instruments such as limit gages, measurement hand tools, gage blocks, and measurement amplifiers.

3.0 DEFINITIONS

3.1   Limit gages: Measurement devices that have fixed sizes and measure the limits of part tolerance on a "go, no-go" basis.

3.2   Measurement hand tools: Portable, hand-operated measurement devices such as micrometers, height gages, depth gages, vernier calipers, and torque wrenches.

3.3   Standards kits: Sets of precision blocks (such as gage blocks) or rods, balls, cylinders, rings, and flats used for checking measurement accuracy.

3.4   Measurement amplifiers: Bench- and pedestal-mounted devices, such as contour projectors and optical height gages, used to amplify measurements to increase accuracy.

4.0 ASSOCIATED MATERIALS

4.1   Calibration System Responsibilities, Quality Procedure 4.2.1

4.2   Equipment Location and Recall Record, Form No. B-23

4.3   Calibration/Service Data Record, Form No. B-24

5.0 PROCEDURE

5.1   All dimensional working instruments are to be certified as to their accuracy on purchase. Certification is the responsibility of the quality department.

5.2   The tool control department of manufacturing has the responsibility for the custody, issuance, and return of dimensional working instruments to tool cribs.

SUBJECT: CALIBRATION AND CONTROL OF DIMENSIONAL WORK- ING INSTRUMENTS		
Authorized By: *V. A. Philis*		**Number:** 4.2.5
Effective Date: 2/10/97	**Revision:** 1/30/97	**Page** 2 **of** ?

     5.3   The tool control department has the responsibility of submitting dimensional hand tools to the quality department for calibration when records indicate that calibration is due.

     5.4   The quality department will create and maintain an Equipment Location and Recall Record (Form No. B-23) and a Calibration/Service Data Record (Form No. B-24) for each dimensional working instrument (see detailed instructions in Quality Procedure 4.2.2).

     5.5   Each dimensional working instrument is to be marked showing the date of the most recent calibration, the stamp of the technician who performed the calibration, and the date when the next calibration is due (see Quality Procedure 4.2.2). If the item is too small for this type of marking, the serial number is to be used for maintaining records that cite the same data.

SUBJECT: SUBCONTRACTING OF CALIBRATION AND REPAIR WORK		
Authorized By: *V. A. Philis*		**Number:** 4.2.6
Effective Date: 2/10/97	**Revision:** 1/30/97	**Page** 1 **of** 3

1.0 PURPOSE

This procedure cites the method to be used at the XYZ company in selecting and monitoring the performance of subcontractors as related to the calibration and repair of measuring and test equipment.

2.0 APPLICATION

Applies to the quality and purchasing department concerning the subcontracting of all measuring and test equipment calibration and repair work.

3.0 ASSOCIATED MATERIALS

3.1   Calibration System Responsibilities, Quality Procedure 4.2.1

3.2   Subcontractor Organization and Facilities Survey and Subcontractor Equipment Survey, Form Nos. B-28 and B-29

4.0 PROCEDURE

4.1   When measuring and test equipment calibration and/or repair work is to be subcontracted, there must be assurances that this work will be done by capable subcontractors and with measurement standards that are certified and traceable to the National Bureau of Standards.

4.2   The responsibility for evaluating candidate calibration/repair subcontractors as to their capabilities to conform to traceability and procedural requirements is that of the measuring equipment control quality department of quality assurance.

4.3   The quality department's evaluation work must be coordinated with the purchasing department, this latter organizational unit responsible for checking candidate subcontractors in regard to other attributes such as financial capabilities.

4.4   In conducting a calibration/repair capabilities survey, the quality department is to prepare a Subcontractor Organization and Facilities Survey Form such as shown in Figure B-28. This form is to be filled out as follows: (1) the potential subcontractor's name, (2) the date of the survey, (3) the organization's location, (4) the name of any subsidiary or affiliate of the organization, (5) the name of the organization's employee who is serving as the chief contact person for this survey, (6) a general description of the organization's calibration and repair capabilities, (7) the number of employees by shift, (8) the square footage of the facility, (9) environmental factors, (10) a statement as to the organization's warranty policy on calibration and repair work, (11) an instruction for the organization to attach to this

SUBJECT: SUBCONTRACTING OF CALIBRATION AND REPAIR WORK		
Authorized By: *V. A. Philis*		**Number:** 4.2.6
Effective Date: 2/10/97	**Revision:** 1/30/97	**Page** 2 **of** 3

survey form a list of on-file calibration instruction manuals and procedures, and (12) the name of two customers as references.

4.5   In conducting a calibration/repair capabilities survey, the quality department is also to prepare a Subcontractor Equipment Survey Form such as the one shown in Figure B-29. This is to be filled out as follows: (1) the potential subcontractor's name, (2) the date of the survey, (3) the calibration/repair equipment items, (4) their manufacturers, (5) their model numbers, and (6) their traceability.

4.6   At the conclusion of each survey, the quality department is to prepare a report that presents the results of the survey and offers a recommendation.

4.7   The quality department is to create and continually maintain a list of qualified subcontractors for measuring and test equipment calibration and repair work.

4.8   The quality department is responsible for maintaining surveillance of selected subcontractors of calibration and repair work, and reporting on their workmanship, their record-keeping performances, and their abilities to comply to schedule requirements.

SUBJECT: SUBCONTRACTING OF CALIBRATION AND REPAIR WORK		
Authorized By: *V. A. Philis*		Number: 4.2.6
Effective Date: 2/10/97	Revision: 1/30/97	Page 3 of 3

**Figure B-28. Subcontractor Organization and Facilities Survey—
Calibration/Repair Capabilities**

Organization (1)	Date (2)
	Contact person (5)

Location (3)   Subsidiary or affiliate of (4)

General description of calibration/repair capabilities (6)

Number of employees (7) 1st shift _____ 2nd shift _____ 3rd shift _____

Size of facility (square footage) (8)	Environmental factors (9) Temperature control range_____
Describe warranty policy on calibration/repair work (10)	Humidity control range_____ Dust control (microns)_____ Other (describe)_____

Attach list of calibration manuals and procedures on file (11)
Customer references (12)

**Figure B-29. Subcontractor Equipment Survey—
Calibration/Repair Capabilities**

Organization (1)			Date (2)
Item (3)	Manufacturer (4)	Model (5)	Traceability (6)

SUBJECT: EMPLOYEE-OWNED MEASURING EQUIPMENT		
Authorized By: *V. A. Philis*		**Number:** 4.2.7
Effective Date: 2/10/97	**Revision:** 1/30/97	**Page** 1 **of** 2

1.0 PURPOSE

To describe the system for controlling the calibration of employee-owned measuring equipment within the XYZ company.

2.0 APPLICATION

Applies to employee-owned measuring devices used in the precision measurement of deliverable products.

3.0 DEFINITION

3.1   Employee-owned measuring equipment: Outside micrometers, inside micrometers, depth gages, vernier calipers, dial indicators, and other similar precision equipment. (Does not include steel rules, feeler gages, radius gages, and other similar items.)

4.0 ASSOCIATED MATERIALS

4.1   Calibration System Responsibilites, Quality Procedure 4.2.1

4.2   Employee-Owned Measuring Equipment Calibration Record, Form No. B-30

5.0 PROCEDURE

5.1   All employee-owned measuring equipment used in the precision measuring of deliverable products must be calibrated by the quality department.

5.2   Employees are required to submit this equipment for calibration via the tool control department's tool cribs.

5.3   The quality department is to create and maintain an Employee-Owned Measuring Equipment Calibration Record for each employee. The form for this, shown in Figure B-30, is filled out as follows: (1) the employee's name, (2) employee's work location, (3) the calibration recall date or dates, (4) the employee's identification number, (5) the name of each tool, (6) the scheduled calibration frequency for each tool, and (7) the actual calibration dates and the initials of the person who performed the calibration.

5.4   Each tool is to be identified with the date of the most recent calibration, the stamp of the technician who performed the calibration, and the date when the next calibration is due (see Quality Procedure 4.2.2). If the item is too small for this type of marking, the serial number is used for maintaining records that cite the same data.

SUBJECT: EMPLOYEE-OWNED MEASURING EQUIPMENT		
Authorized By: *V. A. Philis*		Number: 4.2.7
Effective Date: 2/10/97	Revision: 1/30/97	Page 2 of 2

Figure B-30. Employee-Owned Measuring Equipment Calibration Record

Employee name (1)	Recall (3)
Work location (2)	Employee No. (4)

Tool	Calibration frequency	Calibration dates and/or stamps (7)					
(5)	(6)						

SUBJECT: USE OF INSPECTION EQUIPMENT BY THE CUSTOMER		
Authorized By: *V. A. Philis*	**Number:** 4.3.1	
Effective Date: 2/10/97	**Revision:** 1/30/97	**Page** 1 **of** 2

1.0 PURPOSE

To set forth a method to enable customers to request the use of company measuring and testing devices to verify product conformance with specified requirements within the XYZ company premises.

2.0 APPLICATION

Applies to customers desiring to confirm measuring and/or test results related to products they have contracted to have manufactured.

3.0 PROCEDURE

3.1 Company-owned gages, inspection devices, and test equipment will be made available for use by customers when there is a need to verify product conformance with specified requirements within the XYZ company premises.

3.2 Customers can request the use of these devices using the form shown in Figure B-31. This form is to be filled out by the requester as follows: (1) the name or description of the equipment desired, (2) the equipment identification number, (3) the location where the equipment will be used, (4) the dates when the equipment will be needed, (5) the name of the part or assembly to be measured or tested, (6) the part or assembly number, (7) the contract number, (8) a statement as to the reason for the request, (9) the name of the requester, (10) the requester's title and (11) organization, and (12) the date of the request.

3.3 The customer is to submit the request to the inspection department.

3.4 The inspection department is to evaluate the request, write in space (13) any special conditions, and secure the approvals of management representatives of (14) the manufacturing department and (15) the quality assurance department.

SUBJECT: USE OF INSPECTION EQUIPMENT BY THE CUSTOMER		
Authorized By: *V. A. Philis*		**Number:** 4.3.1
Effective Date: 2/10/97	**Revision:** 1/30/97	**Page** 2 **of** 2

Figure B-31. Customer Request for Use of Measuring/Test Equipment

Equipment required (1)	Equipment No. (2)
Required at following location (3)	Date required (4) From _____ To _____
Part or assembly to be measured/tested (5)	Part or assembly No. (6) Contract No. (7)
Reason for request (8)	

Requested by (9)	Title (10)	Organization (11)	Date (12)

Approved conditions (13)
Approved (manufacturing) (14) _____ Date
Approved (QA) (15)_____ Date

SUBJECT: REQUEST FOR CANDIDATE SUPPLIER QUALITY EVALUATION		
Authorized By: *V. A. Philis*		**Number:** 5.1.1
Effective Date: 2/10/97	**Revision:** 1/30/97	**Page** 1 **of** 3

1.0 PURPOSE

To set forth a procedure that the XYZ company uses for requesting a detailed evaluation of a potential supplier from the standpoint of that supplier's capabilities for producing details or assemblies to specified quality requirements.

2.0 APPLICATION

For use by an organizational unit in requesting that a supplier quality evaluation be made by the quality department of the XYZ company.

3.0 ASSOCIATED MATERIALS

3.1   Candidate Supplier Quality Evaluation Request, Form No. B-32

4.0 PROCEDURE

4.1   Requests to the quality engineering department to conduct a quality evaluation of a potential supplier can be made by any company personnel or organizational unit.

4.2   It is incumbent on the requester to be familiar with the applicable drawings and specifications and reasonably sure that the candidate supplier is capable of meeting these requirements.

4.3   Requests are made either verbally or by filling out the Candidate Supplier Quality Evaluation Request Form, such as shown in Figure B-32. It is to be completed as follows: (1) is preaddressed to Quality Engineering; (2) the date of the request; (3) the name, (4) department, and (5) phone number of the requester; (6) the name, address, and phone number of the candidate supplier; (7) the name of the supplier's parent company, if applicable; (8) the name of the supplier person who should be contacted and (9) that person's title; (10) the Supplier Code Number; (11) an identification description of the item proposed for procurement from this source; (12) a listing of the applicable drawings and specifications; (13) comments, especially relating to justifying this particular source; (14) the dated signature of the requester; and (15) the dated signature of the requester's supervisor.

4.4   On completion of the request, it is to be transmitted to the quality engineering department.

SUBJECT: REQUEST FOR CANDIDATE SUPPLIER QUALITY EVALUATION		
Authorized By: *V. A. Philis*		**Number:** 5.1.1
Effective Date: 2/10/97	**Revision:** 1/30/97	**Page** 2 **of** 3

4.5   The quality engineering department will evaluate the request, acknowledging on the request form itself (16) that the request has been denied, stating the reason for that denial, or (17) scheduling the survey and (18) approving the disposition. A copy of the acknowledged form is to be returned to the requester.

4.6   The purchasing department is charged for maintaining the file for candidate suppliers and evaluation.

SUBJECT: REQUEST FOR CANDIDATE SUPPLIER QUALITY EVALUATION		
Authorized By: *V. A. Philis*		**Number:** 5.1.1
Effective Date: 2/10/97	**Revision:** 1/30/97	**Page** 3 **of** 3

Figure B-32. Candidate Supplier Quality Evaluation Request

TO: Quality Engineering (1)		Date (2)
From (3)	Department (4)	Phone (5)
Supplier (name, address, phone) (6)		Parent company (if applicable) (7)
Contact (8)	Title (9)	Supplier code No. (10)
Item proposed for procurement from this source (11)		
Applicable drawings and/or specifications (12)		
Comments (13)		
Requested by date (14) Date Approved by (15) Date		
Quality engineering acknowledgment _____ Request denied (explain on back or on additional sheet) (16)		
Survey scheduled for (date) (17)		
Approved by (18)		

SUBJECT: QUALITY SURVEY OF CANDIDATE SUPPLIER OR SUBCONTRACTOR		
Authorized By: *V. A. Philis*		**Number:** 5.1.2
Effective Date: 2/10/97	**Revision:** 1/30/97	**Page** 1 **of** 13

1.0 PURPOSE

This procedure describes the XYZ company's method for surveying candidate suppliers and subcontractors to gather information relative to their capabilities for conforming to quality requirements.

2.0 APPLICATION

Two survey forms are described in this procedure: a detailed form for use in gathering information from potential suppliers and subcontractors who are candidates for major levels of procurement activity and a short form for use in gathering information from those who are candidates for minor levels of procurement activity. The conducting of these surveys is the responsibility of the quality engineering department, with recommendations for use by the material department.

3.0 ASSOCIATED MATERIALS

3.1   Request for Candidate Supplier Quality Evaluation, Quality Procedure 5.1.1

3.2   Quality Survey of Candidate Supplier—Short Form, Form No. B-33

3.3   Quality Survey of Candidate Supplier, Form No. B-34

3.4   Candidate Supplier/Subcontractor Quality Evaluation Report, Quality Procedure 5.1.3

4.0 PROCEDURE

4.1   Requests for quality surveys of candidate suppliers or subcontractors can be generated by any company organizational unit (see Quality Procedure 5.1.1).

4.2   Authorization to conduct a quality survey of a candidate supplier or subcontractor can be given by either quality assurance or material department management personnel.

4.3   Surveys are to be conducted by the quality engineering department. If a team is to make the survey, the team members are to be selected from quality engineering (organization team leader), purchasing, project office or program office, and engineering.

4.4   The purchasing department is responsible for making formal survey arrangements with a candidate supplier or subcontractor, including explaining the purpose of the survey and the scheduling of a date and time.

SUBJECT: QUALITY SURVEY OF CANDIDATE SUPPLIER OR SUBCONTRACTOR		
Authorized By: *V. A. Philis*		**Number:** 5.1.2
Effective Date: 2/10/97	**Revision:** 1/30/97	**Page** _2_ **of** _13_

    4.5   A survey is not to begin at a supplier's facility or a subcontractor's facility without an initial, introductory discussion with the appropriate supplier/ subcontractor management representative.

    4.6   At the conclusion of a survey, the supplier/subcontractor management representative is to be verbally briefed as to the findings. Official survey results will later be transmitted by letter.

    4.7   The quality engineering department is to evaluate the survey findings and generate a Candidate Supplier/Subcontractor Quality Evaluation Report (see Quality Procedure 5.1.3).

    4.8   The form to be used in conducting surveys of potential suppliers and subcontractors who may be involved in minor levels of procurement activity or are preapproved by the XYZ's customer is shown in Figure B-33, Quality Survey of Candidate Supplier—Short Form. It is to be filled out as follows: (1) the name, address, and phone number of the supplier; (2) the supplier's general product line; (3) the total number of employees; (4) the number of employees engaged in production; (5) the number of employees in the supplier's quality assurance organization; (6) an identification of the item or items being considered for procurement from this source; (7) a list of government or other specifications compatible with the supplier's quality assurance system; (8) a list of specialized areas (such as clean rooms) and processes (such as plating) that are available in the supplier's facility; (9) whether inspection and test equipment is traceable to the National Bureau of Standards; (10) check marks to indicate which areas are covered by procedures in the supplier's Quality Assurance Manual; (11) remarks (including a notation that the supplier's company and quality assurance organization charts should be appended to this form); (12) the dated signature of the person who made the survey; and (13) the dated signature of the person who approves the survey.

    4.9   Detailed surveys are to be made at the discretion of the quality department of the XYZ company based on the volume and type of business of the vendor using the forms shown in Figure B-34. The cover sheet for this set of forms (Figure B-34, page 1 of 10) is to be filled out as follows: (1) the name, address, and phone number of the candidate supplier; (2) a description of the supplier's primary business and products; (3) the number of employees; (4) an identification of the product being considered from this source; (5) the names of personnel contacted during this survey; (8) the name of his or her department; and (9) the date. The detailed survey sheets (Figure B-34) are to be filled out as follows: (1) the sheet number, (2) the supplier's name, (3) the date, a check mark as to a (4) "yes" or (5) "no" response to each of the listed statements, and (6) an explanation, when required, for each of the "no" responses.

    4.10   When surveys are completed by the appropriate personnel of the XYZ company, they will be placed in the vendor's file.

SUBJECT: QUALITY SURVEY OF CANDIDATE SUPPLIER OR SUBCONTRACTOR		
Authorized By: *V. A. Philis*		**Number:** 5.1.2
Effective Date: 2/10/97	**Revision:** 1/30/97	**Page** 3 **of** 13

Figure B-33. Quality Survey of Candidate Supplier—Short Form

Supplier (name, address, phone) (1)	General product line (2)
	Total No. of employees (3)
	Production employees (4)
	QA employees (5)

Item(s) considered from this source (6)

List quality specs compatible with supplier's QA system (7)

List specialized areas and processes (8)

Is inspection and equipment traceable to national standards? (9)
_____ YES ____ NO

Areas covered by supplier's Quality Assurance Manual (10)

_____ Organization and functions	_____ Records and reports
_____ Inspection stamps	_____ Control of purchases
_____ Nonconforming materials	_____ Material control
_____ Audits	_____ Corrective action
_____ In-process inspection	_____ Final inspection
_____ Training/certification	_____ Drawings and changes
_____ Planning/instructions	_____ Storage delivery

Remarks (11)

Note: Attach copies of company QA organization charts

Prepared by (12) Date	Approved by (13) Date

SUBJECT: QUALITY SURVEY OF CANDIDATE SUPPLIER OR SUBCONTRACTOR		
Authorized By: *V. A. Philis*		**Number:** 5.1.2
Effective Date: 2/10/97	**Revision:** 1/30/97	**Page** 4 **of** 13

Figure B-34. Quality Survey of Candidate Supplier—Cover Sheet Page 1 of 10

Candidate supplier (name, address, phone) (1)
Supplier's primary business/products (2)
Number of employees (3)
Product being considered from this source (4)

Personnel contacted (5)	Title of position (6)

Survey by (7)	Department (8)	Date (9)

SUBJECT: QUALITY SURVEY OF CANDIDATE SUPPLIER OR SUBCONTRACTOR		
Authorized By: *V. A. Philis*	**Number:** 5.1.2	
Effective Date: 2/10/97	**Revision:** 1/30/97	**Page** __5__ **of** __13__

Quality Survey of Candidate Supplier

(1)

Supplier_____ (2) Date _____ (3)   Page 2 of 10

Yes   No*

	Yes	No*
1.0 Quality assurance organization		
1.1 The quality assurance organization's authorities and responsibilities are clearly defined in writing.	(4)	(5)
1.2 The QA organization clearly has the authority to withhold items that have not met acceptable quality standards.		
1.3 The QA organization has direct access to appropriate levels of company management so that quality problems and conflicts can be efficiently and effectively resolved and corrected.		
1.4 There is a Quality Assurance Manual.		
1.5 There is a system for continual maintenance and updating of a Quality Assurance Manual.		
1.6 The quality assurance organization operates a defect prevention program.		
1.7 The quality assurance organization prepares and issues periodic reports and maintains records relative to item acceptance/rejection, disposition of rejected items, quality costs, and other factors.		
1.8 The quality assurance organization maintains a system for the use and control of inspection stamps.		
* Use this space when "no" is indicated and an explanation is required. (6)		

SUBJECT: QUALITY SURVEY OF CANDIDATE SUPPLIER OR SUBCONTRACTOR		
Authorized By: *V. A. Philis*		**Number:** 5.1.2
Effective Date: 2/10/97	**Revision:** 1/30/97	**Page** 6 **of** 13

Quality Survey of Candidate Supplier

(1)

Supplier_____ (2)   Date _____ (3)   Page 3 of 10

	Yes	No*
1.0 Quality assurance organization (cont.)		
1.9 The quality assurance organization indoctrinates and trains employees in the application of quality assurance methods.	(4)	(5)
2.0 Control of procured suppliers		
2.1 The quality assurance organization has a system for quality evaluations of potential suppliers.		
2.2 The QA organization has a program for quality assurance approval of suppliers.		
2.3 The supplier's purchase orders clearly describe the work to be performed.		
2.4 The supplier's purchase orders clearly specify acceptance criteria.		
2.5 The quality assurance organization reviews all purchase orders.		
2.6 The supplier's purchase orders include provisions for customer source inspection and audit.		
2.7 The supplier's purchase orders specify documentation requirements when applicable.		
2.8 The supplier's quality assurance system requires that their own suppliers have adequate quality assurance programs.		
2.9 The quality assurance organization operates a source inspection system.		
* Use this space when "no" is indicated and an explanation is required. (6)		

SUBJECT: QUALITY SURVEY OF CANDIDATE SUPPLIER OR SUBCONTRACTOR		
Authorized By: *V. A. Philis*		**Number:** 5.1.2
Effective Date: 2/10/97	**Revision:** 1/30/97	**Page** 7 **of** 13

Quality Survey of Candidate Supplier

(1)

Supplier_____ (2) Date _____ (3)   Page 4 of 10

                                                            Yes No*

2.0 Control of procured suppliers (cont.)		
2.10 The quality assurance organization operates a source audit program.	(4)	(5)
2.11 The quality assurance organization operates a receiving inspection system.		
2.12 Inspectors are provided with adequate inspection instructions.		
2.13 Receiving inspections records reflect a quality history of suppliers.		
2.14 Drawings used by source inspection and receiving inspection are legible and reflect the latest changes.		
2.15 Source inspectors and receiving inspectors have ready access to the appropriate drawing, engineering orders, specifications, vendor catalogs, purchase orders, and other such materials.		
2.16 The measuring devices, inspection gages, and test equipment available to source inspectors and receiving inspectors are adequate for the inspection and test purchase required.		
2.17 Sampling inspection, when applicable, is performed in compliance with established, recognized standards.		
2.18 The supplier uses a positive means of identification of all raw stock.		

* Use this space when "no" is indicated and an explanation is required. (6)

SUBJECT: QUALITY SURVEY OF CANDIDATE SUPPLIER OR SUBCONTRACTOR		
Authorized By: *V. A. Philis*		**Number:** 5.1.2
Effective Date: 2/10/97	**Revision:** 1/30/97	**Page** 8 **of** 13

Quality Survey of Candidate Supplier

(1)

Supplier_____ (2) Date _____ (3) Page 5 of 10

	Yes	No*
2.0 Control of procured supplies (cont.)		
2.19. The supplier maintains an adequate control area for materials that have been furnished by customers.	(4)	(5)
2.20 The supplier maintains an acceptable system for the age control of items when an item's usability is limited by time.		
2.21 The supplier operates a system that ensures uninspected materials will not be used.		
3.0 In-process inspection		
3.1 An in-process inspection activity is performed, and it is performed by quality assurance.		
3.2 Adequate inspection instructions are made available to all in-process inspection personnel.		
3.3 In-process inspection tasks are performed through the use of written instructions.		
3.4 Drawings used by inspectors are legible and reflect the latest changes.		
3.5 In-process inspectors have ready access to all required drawings, engineering orders, specifications, and other such materials.		
* Use this space when "no" is indicated and an explanation is required. (6)		

SUBJECT: QUALITY SURVEY OF CANDIDATE SUPPLIER OR SUBCONTRACTOR		
Authorized By: *V. A. Philis*		**Number:** 5.1.2
Effective Date: 2/10/97	**Revision:** 1/30/97	**Page** 9 **of** 13

Quality Survey of Candidate Supplier

(1)

Supplier_____ (2) Date _____ (3) Page 6 of 10

Yes No*

3.0 In-process inspection (cont.)		
	(4)	(5)
3.6 The measuring devices, gages, and test equipment required for in-process inspection are available and are adequate.		
3.7 Sampling inspection, when applicable, is performed in compliance with established, recognized standards.		
3.8 The supplier's process capabilities include a certified x-ray laboratory.		
3.9 The supplier's process capabilities include fluorescent and dye-penetrant facilities operated by certified personnel.		
3.10 The supplier's process capabilities include magnetic particle inspection operated by certified personnel.		
3.11 The supplier's process capabilities include resistant welding performed on certified machines.		
3.12 The supplier's process capabilities include fusion welding performed by certified personnel using certified welding materials.		
3.13 The supplier's process capabilities include heat treating performed by qualified personnel using calibrated facilities.		
3.14 The supplier maintains a system that prevents the unauthorized use of materials that have not yet been inspected.		

* Use this space when "no" is indicated and an explanation is required. (6)

SUBJECT: QUALITY SURVEY OF CANDIDATE SUPPLIER OR SUBCONTRACTOR		
Authorized By: *V. A. Phelis*		**Number:** 5.1.2
Effective Date: 2/10/97	**Revision:** 1/30/97	**Page** 10 **of** 13

Quality Survey of Candidate Supplier

(1)

Supplier_____ (2) Date _____ (3) Page 7 of 10

	Yes	No*
3.0 In-process inspection (cont.)		
3.15 The supplier maintains a system for the proper identification of the inspection status of in-process materials.	(4)	(5)
4.0 Final inspection		
4.1 All finished goods are inspected by quality assurance to ensure that contact requirements have been met.		
4.2 Adequate inspection instructions are available to final inspection personnel.		
4.3 Written instructions and procedures are readily available to all final inspection personnel.		
4.4 Drawings used by final inspection personnel are legible and reflect the latest changes.		
4.5 Final inspectors have ready access to all required drawings, engineering orders, specifications, and other materials.		
4.6 The measuring devices, gages, and test equipment required for final inspection are available and adequate.		
4.7 Sampling inspection, when applicable, is performed in compliance with established, recognized standards.		

* Use this space when "no" is indicated and an explanation is required. (6)

SUBJECT: QUALITY SURVEY OF CANDIDATE SUPPLIER OR SUBCONTRACTOR		
Authorized By: *V. A. Philis*	**Number:** 5.1.2	
Effective Date: 2/10/97	**Revision:** 1/30/97	**Page** 11 **of** 13

Quality Survey of Candidate Supplier (1)

Supplier_____ (2) Date _____ (3) Page 8 of 10

	Yes (4)	No* (5)
4.0 Final inspection (cont.)		
4.8 The supplier maintains a system to prevent the unauthorized use of materials that have not yet been inspected.		
5.0 Shipping inspection		
5.1 The supplier's quality assurance organization operates a shipping inspection function.		
5.2 All shipping inspection operations are performed in accordance with written instructions.		
5.3 Shipping inspectors have ready access to customer-specified packaging instructions.		
5.4 When appropriate, packaging tests are conducted.		
5.5 When required, certified packaging materials are used.		
6.0 Measuring devices and test equipment		
6.1 The quality assurance organization maintains procedures that call for the periodic inspection and recalibration of all measuring devices, gages, and items of test equipment.		
6.2 The quality assurance organization maintains procedures that call for the periodic inspection of all production tools used as a medium of inspection in the production processes.		

* Use this space when "no" is indicated and an explanation is required. (6)

SUBJECT: QUALITY SURVEY OF CANDIDATE SUPPLIER OR SUBCONTRACTOR		
Authorized By: *V. A. Philis*	**Number:** 5.1.2	
Effective Date: 2/10/97	**Revision:** 1/30/97	**Page** 12 **of** 13

Quality Survey of Candidate Supplier

(1)

Supplier_____ (2) Date _____ (3) Page 9 of 10

Yes No*

6.0 Measuring devices and test equipment (cont.)		
6.3 The supplier maintains working standards of required accuracy that are periodically calibrated to primary standards traceable to the National Bureau of Standards.	(4)	(5)
6.4 The supplier's quality assurance organization maintains a system for periodically calibrating the tools and gages that are owncd by employees.		
6.5 Whenever measuring devices, gages, or test equipment items are reworked, they are inspected and calibrated prior to use.		
6.6 When new measuring devices, gages, and test equipment are acquired, they are inspected and calibrated prior to use.		
6.7 The processes for calibrating measuring devices, gages, and test equipment are covered by written procedures.		
6.8 All measuring devices, gages, and test equipment items carry stamps that indicate the most recent calibration date and the date when the next calibration is to be performed.		
6.9 The supplicr's quality assurance organization maintains a system for the automatic recall and periodic recalibration of all measuring devices, gages, and test equipment.		
* Use this space when "no" is indicated and an explanation is required. (6)		

SUBJECT: QUALITY SURVEY OF CANDIDATE SUPPLIER OR SUBCONTRACTOR		
Authorized By: *V. A. Philis*		**Number:** 5.1.2
Effective Date: 2/10/97	**Revision:** 1/30/97	**Page** 13 **of** 13

Quality Survey of Candidate Supplier

(1)

Supplier_____ (2) Date _____ (3)   Page 10 of 10

                                                              Yes No*

	Yes (4)	No* (5)
7.0 Material review		
7.1 The supplier maintains a documented system for the handling of nonconforming materials.		
7.2 The supplier maintains a system for removing nonconforming supplies from the production flow.		
7.3 The supplier maintains a system for taking corrective action in order to prevent repetitive discrepancies.		
7.4 The supplier's correction action system is one that permits prompt, remedial actions.		
7.5 The supplier maintains a system for following up on all corrective action requests.		
7.6 Reports on nonconforming materials are regularly prepared and are reviewed by management for action.		

* Use this space when "no" is indicated and an explanation is required. (6)

SUBJECT: CANDIDATE SUPPLIER/SUBCONTRACTOR QUALITY EVALUATION REPORT		
Authorized By: *V. A. Philis*	**Number:** 5.1.3	
Effective Date: 2/10/97	**Revision:** 1/30/97	**Page** 1 **of** 2

1.0 PURPOSE

To provide the XYZ company with a procedure for the reporting of candidate supplier/subcontractor survey results.

2.0 APPLICATION

Applies to all Candidate Supplier/Subcontractor Surveys. The report preparation is the responsibility of the quality engineering department.

3.0 ASSOCIATED MATERIALS

3.1 Quality Survey of Candiate Supplier—Short Form, Form No. B-33

3.2 Quality Survey of Candidate Supplier, Form No. B-34

3.3 Candidate Supplier/Subcontractor Quality Evaluation Report, Form No. B-35

4.0 PROCEDURES

4.1 All Quality Surveys of Candidate Suppliers and Subcontractors (see Quality Procedure 5.1.2) are to be reviewed and evaluated by quality engineering.

4.2 In the case of candidate suppliers and subcontractors who have performed work for the company in the past, their historical quality ratings are to be procured and studied (see Quality Procedure 5.2.3).

4.3 Each evaluation is to be concluded with one of three recommended dispositions: approval, disapproval, or conditional approval. The conditional approval is given on the basis that the prospective supplier or subcontractor will take corrective action to eliminate problems or discrepancies.

4.4 The report is to be prepared on the form shown in Figure B-35, Candidate Supplier/Subcontractor Quality Evaluation Report. It is to be filled out as follows: (1) the name, address, and phone number of the supplier; (2) an identification of the type of survey that was made; (3) a description of the item or items being considered for procurement from this source; (4) the conclusions of the evaluation; (5) a check mark to indicate the recommended disposition as approved, disapproved, or conditionally approved; (6) an identification of the conditions that must be met if the potential supplier/subcontractor was conditionally approved; (7) the dated signature and title of the person who made the evaluation; and (8) the dated signature and title of the person who approved the evaluation.

4.5 The report, with the survey attached, is to be forwarded to the purchasing department for filing in the vendor's file.

SUBJECT: CANDIDATE SUPPLIER/SUBCONTRACTOR QUALITY EVALUATION REPORT	
Authorized By: *V. A. Philis*	**Number:** 5.1.3
Effective Date: 2/10/97 **Revision:** 1/30/97	**Page** 2 **of** 2

Figure B-35. Candidate Supplier/Subcontractor Quality Evaluation Report

Supplier (name, address, phone) (1)	Type of survey (2)
	____ Survey—short form
	____ Detailed survey
	____ ISO 9000
	____ QS-9000
	____ Historical ratings (attach copy)

Item(s) considered from this source (3)

Evaluation (4)

Recommended disposition (5)
____ Approved ____ Disapproved ____ Conditionally approved

Conditions (6)

Evaluation by (name and title) (7)	Date
Approved by (name and title) (8)	Date

SUBJECT: SUPPLIER PROCESS CERTIFICATION		
Authorized By: *V. A. Philis*		**Number:** 5.1.4
Effective Date: 2/10/97	**Revision:** 1/30/97	**Page** __1__ **of** __3__

1.0 PURPOSE

To provide the XYZ company with a system for certifying specific supplier processes when such certification is required.

2.0 APPLICATION

Applies to special or unique quality-sensitive processes performed by suppliers.

3.0 ASSOCIATED MATERIALS

3.1   Supplier Process Certification, Form No. B-36

4.0 PROCEDURE

4.1   Requests for the certification of a supplier's process can be originated, in memo form, by the engineering, material, manufacturing, or quality assurance departments.

4.2   Requests for such certification are to be directed to the quality engineering department.

4.3   On receipt of a certification request, the quality engineering department is to visit the supplier and conduct a detailed analysis of the process or processes in question.

4.4   If the supplier's process or processes cannot be certified, the quality engineering department prepares a memo to this effect and transmits it to the original requester.

4.5   If the supplier's process or processes can be certified, the form shown in Figure B-36 is filled out as follows: (1) the certification number; (2) the certification date; (3) the certification's expiration date (normally one year after certification); (4) the name, address, and phone number of the supplier; (5) the date when the survey was made; (6) the name and/or description of the process or processes being certified; (7) the specifications to which the processes comply; (8) the dated signature of the person who has prepared the certification form; and (9) the dated signature of the person who approves the certification.

4.6   The certification is to be forwarded to the material's purchasing department, where the master record is to be maintained.

4.7   A copy of the certification is to be transmitted to the requester.

SUBJECT: SUPPLIER PROCESS CERTIFICATION		
Authorized By: *V. A. Philis*		Number: 5.1.4
Effective Date: 2/10/97	Revision: 1/30/97	Page 2 of 3

4.8   A copy of the certification is to be maintained by expiration date in a tickler file in the quality engineering department.

4.9   The certification file is to be continually reviewed by quality engineering for the purpose of activating a recertification investigation of suppliers with certifications approaching their expiration dates.

SUBJECT: SUPPLIER PROCESS CERTIFICATION		
Authorized By: *V. A. Philis*		**Number:** 5.1.4
Effective Date: 2/10/97	**Revision:** 1/30/97	**Page** 3 **of** 3

Figure B-36. Supplier Process Certification

Certification No. (1)	Certification date (2)	Expiration date (3)

Supplier (name, address, phone) (4)	This supplier, surveyed _____ (5) has been qualified to perform work utilizing the processes specified as follows:
Processes	Specifications
(6)	(7)
Prepared by (8) Date	
Approved by (9) Date	

SUBJECT: PROCUREMENT DOCUMENT QUALITY REQUIREMENT REVIEW		
Authorized By: *V. A. Philis*		Number: 5.1.5
Effective Date: 2/10/97	Revision: 1/30/97	Page 1 of 2

1.0 PURPOSE

To set forth a method for reviewing all procurement documents, prior to their issuance, to ensure that contractual and internal quality requirements have been properly and completely specified.

2.0 APPLICATION

This procedure applies to the procurement of all materials, details, and assemblies that will be used in deliverables to the XYZ company's customers. It involves specific actions of both the material and quality engineering functions.

3.0 ASSOCIATED MATERIALS

3.1   Purchase order

3.2   Purchase Order Change Notices

3.3   Subcontracts

3.4   Inspection/Test Instructions, Quality Procedure 3.3.1

4.0 PROCEDURE

4.1   Prior to issuance, the purchasing department is to make available all purchase orders, Purchase Order Change Notices, and subcontracts to the quality department for review when the purchasing department has been specifically requested to do so, either verbally or by the requisition form.

4.2   Quality engineering is to review these procurement documents to determine if all contractual and internal quality requirements have been adequately specified.

4.3   Each procurement document is to be augmented with the appropriate Inspection/Test Instruction.

4.4   Each procurement document is to specify that company procurement personnel, quality assurance personnel, and engineering personnel will have access to the supplier's or subcontractor's premises.

4.5   If test reports and/or certifications are to be provided by the supplier or subcontractor, those items are to be specified in the procurement document.

SUBJECT: PROCUREMENT DOCUMENT QUALITY REQUIREMENT REVIEW		
Authorized By: *V. A. Philis*	**Number:** 5.1.5	
Effective Date: 2/10/97	**Revision:** 1/30/97	**Page** 2 **of** 2

4.6   If the supplier or subcontractor is to maintain inspection or test records or other records as to functional, chemical, and/or physical properties, this fact is to be specified in the procurement document.

4.7   If there are discrepancies in the procurement document, it is to be returned to the material department accompanied by a memo that describes the problem.

4.8   If the procurement document is determined to be adequate from a quality requirements' standpoint, it is to be initialed as approved by the quality engineering manager and returned to the purchasing department.

SUBJECT: SOURCE INSPECTION		
Authorized By: *V. A. Philis*		**Number:** 5.2.1
Effective Date: 2/10/97	**Revision:** 1/30/97	**Page** <u>1</u> **of** <u>2</u>

1.0 PURPOSE

This procedure describes the XYZ company's approach for the source inspection of parts or assemblies produced by suppliers and subcontractors.

2.0 APPLICATION

This procedure applies to all suppliers and subcontractors, and its application is the responsibility of the inspection and test department. It includes the functions of both in-process inspection and final inspection.

3.0 ASSOCIATED MATERIALS

3.1   Source Inspection Status Report, Form No. B-37

4.0 PROCEDURE

4.1   Source inspection of details or assemblies at a supplier's facility or subcontractor's facility will be performed whenever it is specified as a requirement in a contract or purchase order.

4.2   The Source Inspection Status Report, Figure B-37, is filled out in the following manner: (1) the name and address of the vendor, (2) the number of the report, (3) the date, (4) a check mark to indicate if action is required or if the purpose of the report is to serve as information only, (5) the name of the program or project, (6) the item to be inspected, (7) the specification number, (8) the purchase order number, (9) the engineering status, (10) the material status, (11) the schedule status, (12) the quality status, (13) the dated signature of the person making the inspection, and (14) the dated signature of the person reviewing the report.

4.3   The source inspection is made at the point of fabrication and assembly prior to shipment to this company. It is made using inspection instructions, drawings, and specifications and includes an examination of the supplier's or subcontractor's records of inspections and tests. The results are recorded on the Source Inspection Status Report.

SUBJECT: SOURCE INSPECTION		
Authorized By: *V. A. Philis*		**Number:** 5.2.1
Effective Date: 2/10/97	**Revision:** 1/30/97	**Page** 2 **of** 2

Figure B-37. Source Inspection Status Report

Supplier (name, location) (1)	Report No. (2)
	Date (3)
	Action required (4) ____ Yes ____ No Information only ____ Yes ____ No
Program or project (5)	
Item (6)	

Spec. No. (7)	P.O. No. (8)

Engineering status (9)	
Material status (10)	
Schedule status (11)	
Quality status (12)	

Inspected by (13) Date	Renewed by (14) Date

SUBJECT: CONTRACTED INSPECTION		
Authorized By: *V. A. Philis*		**Number:** 5.2.2
Effective Date: 2/10/97	**Revision:** 1/30/97	**Page** 1 **of** 2

1.0 PURPOSE

The purpose of this procedure is for the XYZ company to establish a method for obtaining and using the services of a source inspection representative when circumstances dictate the need for this type of service.

2.0 APPLICATION

2.1 The circumstances under which the use of a source inspection representative might be considered are as follows: the source inspection workload will exceed, on a temporary basis, the source and receiving inspection department, or the distance between the supplier or subcontractor, and this company is of such a magnitude that it would be more economical to engage the services of a source inspection service in the source's immediate locality.

2.2   Services that can be covered by the representative include inspection, test witnessing, and periodic audits. Control of the representative's activities will be under the jurisdiction of the source and receiving inspection department.

3.0 PROCEDURE

3.1   When it has been determined that there is a potential need for contracting an inspection service, the quality department will prepare a report identifying the supplier or subcontractor in terms of location, types and quantities of items being fabricated or assembled, and date or dates when source inspection tasks are to be performed.

3.2   The quality department is to identify the type or types of tasks to be performed: inspection, witnessing of tests, and/or periodic audits.

3.3   The project manager is next required to justify the need for contracting for source inspection services by describing the requirement, explaining the reason for not utilizing existing department resources to perform the work, describing the benefits of the intended subcontracting approach, and estimating the cost of the intended subcontracted effort.

3.4   The report requesting and justifying the use of a contracted source inspection representative is to be submitted to the vice president of administration for review and approval.

3.5   On approval, the quality department is to coordinate with the project manager in evaluating and selecting a source inspection representative.

SUBJECT: CONTRACTED INSPECTION		
Authorized By: *V. A. Philis*		Number: 5.2.2
Effective Date: 2/10/97	Revision: 1/30/97	Page _2_ of _2_

3.6   The purchase order or contract prepared for source inspection services requires the approval of both the quality assurance and project manager.

3.7   The issuing of work instructions to the source inspection service organization and the monitoring of its work will be the responsibility of the quality department.

SUBJECT: SUPPLIER QUALITY RATING		
Authorized By: *V. A. Philis*		Number: 5.2.3
Effective Date: 2/10/97	Revision: 1/30/97	Page 1 of 3

1.0 PURPOSE

To define the method that the XYZ company uses for rating the quality and workmanship performance of suppliers and subcontractors.

2.0 APPLICATION

Requires the evaluation, rating, and reporting by the receiving final inspection department of the quality and workmanship performance of all suppliers and subcontractors.

3.0 ASSOCIATED MATERIALS

3.1   Receiving Inspection Report, Form No. B-40

3.2   Supplier/Subcontractor Monthly Quality and Workmanship Report, Form No. B-38

3.3   The XYZ company's Inspection Report on a Pareto chart

3.4   Corrective Action Report

4.0 PROCEDURE

4.1   The evaluating and rating of supplier/subcontractor performance in terms of quality and workmanship are the responsibility of the receiving final inspection department.

4.2   Evaluations and ratings are to be made by the inspection department.

4.3   Source materials to be used in evaluating and rating supplier/subcontractor performance include, but are not limited to, Receiving Inspection Reports (Form No. B-40).

4.4   The form to be used in the monthly evaluation and rating of each supplier/subcontractor is shown in Figure B-38, Supplier/Subcontractor Monthly Quality and Workmanship Report. It is to be filled out as follows: (1) the name of the supplier or subcontractor; (2) the month covered by the report; (3) the name of the program or project; (4) the purchase order or contract number; (5) the part or assembly number and each process (such as welding, painting, assembly, packaging, etc.) used by the supplier or subcontractor; (6) a listing of each process; (7) a check mark indicating whether the work connected with that process was performed in a poor, fair, or good manner; (8) remarks relative to each process; (9) general remarks; percentage ratings in terms of (10) quality

SUBJECT: SUPPLIER QUALITY RATING		
Authorized By: *V. A. Philis*	**Number:** 5.2.3	
Effective Date: 2/10/97	**Revision:** 1/30/97	**Page** 2 **of** 3

remarks and (11) delivery remarks; (12) the dated signature of the person who has prepared the report; and (13) the dated signature of the person approving the report.

4.5   When initiated, copies of the Supplier/Subcontractor Monthly Quality and Workmanship Report are to be distributed to the quality department and the purchasing department.

4.6   The purchasing manager is responsible for limiting or suspending the use of suppliers or subcontractors who consistently have quality or delivery deficiencies.

SUBJECT: SUPPLIER QUALITY RATING		
Authorized By: _V. A. Philis_		**Number:** 5.2.3
Effective Date: 2/10/97	**Revision:** 1/30/97	**Page** 3 **of** 3

Figure B-38. Supplier/Subcontractor Monthly Quality and Workmanship Report

Supplier/subcontractor (1)				Covering month of (2) _____ 19____
Project or program (3)				P.O. or contract No. (4)
Part or assembly (5)				
Process (6)	Rating (check) (7)			Remarks (8)
	Poor	Fair	Good	
General remarks (9)				
Quality (10)				
Delivery (11)				
Prepared by (12) Date			Approved by (13) Date	

SUBJECT: SUPPLIER PART QUALIFICATION		
Authorized By: *V. A. Philis*	**Number:** 6.1.1	
Effective Date: 2/10/97 **Revision:** 1/30/97	**Page** __1__ **of** __1__	

1.0 PURPOSE

To set forth the method that the XYZ company uses for processing and evaluating details and related test reports submitted by candidate suppliers in order to qualify those details for use in company products.

2.0 APPLICATION

Applies to the qualification of all details submitted to the company for evaluation and for which the company has determined a need.

3.0 PROCEDURE

3.1   Company requests to candidate suppliers for parts and data to be submitted for qualification purposes will be made through the use of purchase orders prepared and issued by the purchasing department.

3.2   Submitted details and test reports are to be routed through receiving inspection to the final inspection.

3.3   Receiving inspection will examine the incoming materials for transit damage and completeness, coordinating the correction of deficiencies in these areas prior to acceptance.

3.4   When discrepancies are found, copies of the report (Form B-40) are to be distributed to the material control department.

SUBJECT: FIRST ARTICLE INSPECTION—PROCURED MATERIAL		
Authorized By: *V. A. Philis*		**Number:** 6.1.2
Effective Date: 2/10/97	**Revision:** 1/30/97	**Page** 1 **of** 3

1.0 PURPOSE

To define a method that the XYZ company uses for inspecting the first of a series of articles produced by a subcontractor for the purpose of early detection and timely correction of discrepancies.

2.0 APPLICATION

Applies to the first of all new or significantly changed nonstandard parts and assemblies manufactured for the company by subcontractors or the first nonstandard part or assembly manufactured for the company by a new subcontractor. The first article inspection of internally manufactured parts and assemblies is covered by Quality Procedure 6.3.1.

3.0 ASSOCIATED MATERIAL

3.1   First Article Report—Procured Material, Form No. B-39

4.0 PROCEDURE

4.1   The purchasing department is responsible for citing on a purchase order or contract the requirement for a first article inspection.

4.2   First article inspection normally will be performed in the receiving area. The inspection can be performed by the source and an inspection report submitted to the quality department for review and acceptance if so authorized by the quality department.

4.3   The first article inspection is to be a complete one, covering all appropriate physical and functional characteristics of the part or assembly, workmanship, and the completeness and correctness of the required documentation.

4.4   The quality department is responsible for provisioning the appropriate department with the required test equipment, test equipment operating instructions, appropriate drawings, and/or specifications.

4.5   At the completion of the first article inspection, the inspector is to prepare the First Article Report—Procured Material. A copy of the form is shown in Figure B-39. The form is to be filled out as follows: (1) the name and address of the subcontractor; (2) the date of the report; (3) the report number; (4) the part or assembly number; (5) the revision number; (6) the part or assembly name; (7) the next assembly number; (8) the purchase order or contract number; a check mark either at (9) indicating approval or at (10) indicating rejection; and, in the case of rejection (11), a description of the discrepancy; (12) recommended disposition;

SUBJECT: FIRST ARTICLE INSPECTION—PROCURED MATERIAL		
Authorized By: *V. A. Philis*		Number: 6.1.2
Effective Date: 2/10/97	Revision: 1/30/97	Page _2_ of _3_

(13) a check mark to indicate whether or not a test report is attached; (14) the dated signature of the inspector; and (15) the dated signature of the person authorized to approve the report.

4.6   Copies of the report are to be distributed as follows: receiving inspection files, purchasing (for forwarding to the subcontractor), quality department, and project manager.

SUBJECT: FIRST ARTICLE INSPECTION—PROCURED MATERIAL		
Authorized By: *V. A. Philis*		**Number:** 6.1.2
Effective Date: 2/10/97	**Revision:** 1/30/97	**Page** 3 **of** 3

Figure B-39. First Article Report—Procured Material

Subcontractor (name and address) (1)	Date (2)	Report No. (3)
	Part/assembly (4)	Revision (5)
	Name (6)	
Next assembly No. (7)	Purchase order or contract No. (8)	
First article approved ____ (9) This is your authority to proceed First article rejected ____ (10) Correct defects and resubmit sample		
Discrepancy (11)		
Recommended disposition (12)		
Detailed test report attached? (13) ____ Yes ____ No		
Inspected by (14) Date	Approved by (15) Date	

SUBJECT: RECEIVING INSPECTION		
Authorized By: *V. A. Philis*		Number: 6.1.3
Effective Date: 2/10/97	Revision: 1/30/97	Page 1 of 4

1.0 PURPOSE

To prescribe the system that the XYZ company uses for inspecting materials received from suppliers and subcontractors.

2.0 APPLICATIONS

Applies to the inspection of all materials received except raw materials, the latter covered by Quality Procedure 6.1.4.

3.0 ASSOCIATED MATERIALS

3.1   Receiving Inspection Report, Form No. B-40

3.2   Material Return Notice, Form No. B-41

3.3   Rejection and Disposition Tag, Form No. B-49

4.0 PROCEDURE

4.1   All materials that are to be inspected on receipt are to be inspected by receiving inspection in terms of compliance to specified requirements, completeness, transit damage, and proper and complete paperwork and documentation.

4.2   Parts that have been sent out for special processing are to be inspected, when returned, only for the processing performed.

4.3   Measuring and test equipment devices and measurement standards that have been sent out for calibration and repair, when returned, are to be forwarded to the quality department for inspection.

4.4   All incoming materials are to be processed in the priority sequence of the dates when the materials are required.

4.5   The materials are to be inspected for conformance to requirements, recording results on the Receiving Inspection Report Form, shown in Figure B-40. The data to be recorded are as follows: (1) the name of the vendor, supplier, or subcontractor; (2) the purchase order or contract number; (3) the invoice number or shipper number; (4) the report date; and for each separate type of item, (5) the item number, (6) the item description, (7) the inspection criteria, (8) the quantity ordered, (9) the quantity cited on the packing slip, (10) the quantity received, (11) the quantity inspected, (12) the quantity accepted, and (13) the quantity rejected.

SUBJECT: RECEIVING INSPECTION		
Authorized By: *V. A. Philis*		Number: 6.1.3
Effective Date: 2/10/97	Revision: 1/30/97	Page 2 of 4

Descriptions of the discrepancies found are recorded as follows: (14) handling/ shipping defects, (15) discrepancy between quantities shipped and received, (16) improper paperwork, (17) incomplete and/or incorrect documentation, (18) discrepancies regarding physical characteristics, (19) discrepancies regarding functional characteristics, and (20) other problems. The inspector's stamp or initials on the report in the lower, right-hand corner of the form (21).

4.6   Materials that have been rejected should be tagged with the Rejection and Disposition Tag, Form No. B-49 (see Quality Procedure 6.5.3), and forwarded to the material review board for disposition or returned directly to the vendor, supplier, or subcontractor.

4.7   For materials to be returned, the Material Return Notice Form should be prepared. An example of this form is displayed in Figure B-41. This should be filled out as follows: (1) the name and address of the vendor, supplier, or subcontractor to whom the materials are being returned; (2) a check mark indicating that the materials are being returned either for credit only or for rework or replacement; (3) the name and address of the sender; (4) the date of this notice; (5) the debit number; (6) the purchase order or contract number; (7) the invoice number; (8) the name of the inspector; (9) the name of the buyer; (10) the signature of the person authorized to sign this notice; (11) the quantity of materials being returned; (12) the part numbers; (13) the description; (14) the reason for the rejection; and (15) a check mark to indicate if and when a corrective action statement would be required.

SUBJECT: RECEIVING INSPECTION		
Authorized By: *V. A. Phtis*	**Number:** 6.1.3	
Effective Date: 2/10/97	**Revision:** 1/30/97	**Page** 3 **of** 4

Figure B-40. Receiving Inspection Report

Supplier/subcontractor (1)					P.O. or contract No. (2)			
					Invoice No. (3)			
					Report date (4)			
Item No.	Item description	Inspection criteria	Quantities					
			Ordered	Pack, slip	Received	Inspected	Accepted	Rejected
(5)	(6)	(7)	(8)	(9)	(10)	(11)	(12)	(13)

Description of discrepancies (use extra sheets if necessary)
Handling/shipping defect (14)
Discrepancy between quantities shipped and received (15)
Improper paperwork (16)
Incomplete/incorrect documentation (17)
Discrepancies regarding physical characteristics (18)
Discrepancies regarding functional characteristics (19)
Other problems (20)
Inspected by (21)

SUBJECT: RECEIVING INSPECTION		
Authorized By: *V. A. Philis*		**Number:** 6.1.3
Effective Date: 2/10/97	**Revision:** 1/30/97	**Page** 4 **of** 4

Figure B-41. Material Return Notice

Return to (supplier or subcontractor) (1)	Rejected material (2) Itemized material is returned _____ For credit only—do not replace _____ For rework or replacement Your account is being debited accordingly, reinvoice if material is reworked or replaced	
Returned by (3)	Date (4)	Debit No. (5)
	P.O. or contract No. (6)	Invoice No. (7)
	Inspector (8)	Buyer (9)
	Authorized by (10)	

Quantity	Part No.	Description	Reason for rejection
(11)	(12)	(13)	(14)

NOTICE TO SUPPLIER	Please send a corrective action statement (15) _____ With next statement _____ Within _____ days

SUBJECT: RAW MATERIAL INSPECTION		
Authorized By: *V. A. Philis*		**Number:** 6.1.4
Effective Date: 2/10/97	**Revision:** 1/30/97	**Page** 1 **of** 2

1.0 PURPOSE

To establish the requirements that the XYZ company uses for inspecting raw materials that have been received.

2.0 APPLICATION

Applies to the receiving inspection of raw materials. The receiving inspection of all other materials is covered by Quality Procedure 6.1.3.

3.0 DEFINITION

3.1   Raw material: Basic materials from which articles are manufactured. Raw materials include sheet metal, metal bar stock, metal and plastic extrusions, pipe stock, wire, plastics, rubber materials, chemicals, and other materials.

4.0 ASSOCIATED MATERIALS

4.1   Receiving Inspection Report, Form No. B-40

4.2   Rejection and Disposition Tag, Form No. B-49

5.0 PROCEDURE

5.1   The purchasing department is to specify, on all purchase orders for raw materials, specifications as to the physical and/or chemical characteristics and properties that must be in compliance.

5.2   When deemed appropriate by the purchasing department and quality assurance, the purchase order is to specify that the supplier provide certificates and test reports attesting to the raw material's compliance to specified requirements.

5.3   Receiving inspection is responsible for inspecting incoming raw materials against the purchase order and other applicable documents.

5.4   At the completion of inspection (if a discrepancy exists), receiving inspection will prepare a Receiving Inspection Report, Form No. B-40 (see Quality Procedure 6.1.3). If the incoming materials are steel bars, then receiving marks them according to codes and colors in the following manner:

SUBJECT: RAW MATERIAL INSPECTION		
Authorized By: *V. A. Philis*	**Number:** 6.1.4	
Effective Date: 2/10/97	**Revision:** 1/30/97	**Page** 2 **of** 2

Steel Inventory Colors

CRS	Orange
8620	Yellow
6150	White
1060	Blue
1045	Black
0-1	Red
4140	Green
4140-HT	Dark green
A-2	Gold
W-2	Silver
4150	Gray
HRS	Brown

5.5   Unacceptable materials are to be tagged with the Rejection and Disposition Tag (Form No. B-49) and turned over to the receiving and shipping department for disposition at the request of the material control department.

SUBJECT: IN-PROCESS INSPECTION		
Authorized By: *V. A. Philis*		Number: 6.2.1
Effective Date: 2/10/97	Revision: 1/30/97	Page 1 of 1

1.0 PURPOSE

This procedure establishes the method that the XYZ company uses for monitoring and controlling the quality of parts, components, and subassemblies throughout the various intermediate steps involved in the manufacturing process.

2.0 APPLICATION

Applies to the in-process inspection department of quality assurance and excludes source and receiving inspection and final inspection and test.

3.0 ASSOCIATED MATERIALS

3.1   Drawings as required for specific job.

4.0 PROCEDURE

4.1   Inspection methods employed can include inspections by machine operators and may be witnessed by inspection personnel. Setup or first piece approval is at the discretion of the department manager.

4.2   Inspections are to be made using applicable inspection instructions, drawings, specifications, and other appropriate reference materials.

4.3   The inspection is to include an examination of the drawings and specifications for completeness and correctness, workmanship, physical and functional characteristics, and, when called for, the effectiveness of special processes such as heat treating or the application of special coatings.

4.4   Materials to be reworked on the spot. If not repairable, the leader is notified and disposition to scrap material is made or sent to the appropriate department together with rework instructions.

SUBJECT: HEAT TREAT INSPECTION		
Authorized By: *V. A. Philis*		**Number:** 6.2.2
Effective Date: 2/10/97	**Revision:** 1/30/97	**Page** 1 **of** 1

1.0 PURPOSE

To describe the method and responsibilities that the XYZ company uses for inspecting parts and materials that have been heat treated.

2.0 APPLICATION

Applies to the inspection of all deliverable carbon and alloy steel parts and materials that have been hardened, tempered, and/or stress relieved.

3.0 ASSOCIATED MATERIALS

3.1   Forms supplied by the heat treating service.

4.0 PROCEDURE

4.1   The requirement for heat treat inspection will be specified by the leader of the specific job, and those instructions will be transferred to the form supplied by the heat treating service.

4.2   Hardness tests will be made utilizing either Rockwell or Brinell hardness equipment and procedures.

4.3   On completion of the tests, parts that meet drawing specifications are to be identified as to hardness and can continue in their normal flow through the manufacturing process.

4.4   Discrepancies are to be referred to the shop superintendent.

SUBJECT: WELDING AND BRAZING INSPECTION		
Authorized By: *V. A. Philis*	**Number:** 6.2.3	
Effective Date: 2/10/97	**Revision:** 1/30/97	**Page** 1 **of** 1

1.0 PURPOSE

To describe the method and responsibilities for inspecting parts and materials that have been welded or brazed.

2.0 APPLICATION

Applies to all welded or brazed deliverable parts and materials, the purpose being to ensure that permanent joints have been obtained between connecting pieces.

3.0 PROCEDURE

3.1   Welders are to be certified per Specification AWS-D1.1 and AWS-B3-77.

SUBJECT: FAILURE REPORTING		
Authorized By: *V. A. Philis*		**Number:** 6.2.4
Effective Date: 2/10/97	**Revision:** 1/30/97	**Page __1__ of __2__**

1.0 PURPOSE

To set forth the methods and responsibilities that the XYZ company uses for initiating and issuing a failure report and for specifying and reporting on actions to correct the problem during the debug run or any other tryout.

2.0 APPLICATION

To be used when failures are detected in units, subsystems, or systems.

3.0 ASSOCIATED MATERIALS

3.1   Failure Report, Form No. B-42

4.0 PROCEDURE

4.1   A failure report is to be initiated by the tryout leader. Failures detected, including those discovered during unit test, system test, acceptance test, and field test, are recorded on the Failure Report.

4.2   The form to be used for this purpose is displayed in Figure B-42. It is to be filled out as follows: (1) the report number, (2) the system or assembly name, (3) the date of the failure, (4) the system or assembly number, (5) the serial number, (6) the amount of time the system was run during testing, (7) failure description and apparent cause, (8) when the failure was detected (unit test, system test, acceptance test, field test, or other), (9) the environmental conditions at the time of the test, (10) the name of the person preparing the report, (11) that person's department, and (12) repair instructions.

4.3   The failure report is to accompany the system or assembly to the area where the rework is to be performed. Repair information is to be added to the report at that point in terms of (13) repair notes; actual personnel hours used for (14) fault isolation and (15) replacement and/or repair; information as to parts removed, including (17) part number, (18) serial numbers, (19) manufacturers, and (20) defects; and (21) the serial numbers of the replacement parts. The name of the technician who made the repairs is given in space (22).

4.4   After rework, the system or assembly is to be submitted for retest to verify that the failure has been corrected. Retest data is recorded on the Failure Report in space (16) actual personnel hours for retesting and reinspection, (23) the name of the person who conducted the retest, and (24) an indication as to whether or not the trouble was corrected.

4.5   Removed items are to be tagged with the Rejection and Disposition Tag (Form No. B-49) and routed to the material review board for disposition.

SUBJECT: FAILURE REPORTING		
Authorized By: *V. A. Philis*	**Number:** 6.2.4	
Effective Date: 2/10/97	**Revision:** 1/30/97	**Page** 2 **of** 2

Figure B-42. Failure Report

System or assembly name (2)	Report No. (1)
	Failure date (3)
	Serial No. (5)
System or assembly No. (4)	System run time (6)

Failure description and apparent cause (7)

When detected (8) ___ Unit test ____ System test _____ Accepted test
___ Field test ____ Other (specify)

Environmental control (9)   Temperature _____   Humidity _____
                            Other _____

Report prepared by (10)	Department (11)

Repair instructions (12)

Repair notes (13)	Actual personnel hours Fault isolation (14) ____ Replace/repair (15) ____ Retest/reinspect (16) ____

Parts removed				Replaced by serial No.
Number	Ser. No.	Manufacturer	Defect	
(17)	(18)	(19)	(20)	(21)

Repaired by (22)	Retested by (23)	Trouble corrected (24) Yes _____ No _____

SUBJECT: TOOLING INSPECTION		
Authorized By: *V. A. Philis*		**Number:** 6.2.5
Effective Date: 2/10/97	**Revision:** 1/30/97	**Page** 1 **of** 3

1.0 PURPOSE

To establish the method that the XYZ company uses for inspecting production tools, including jigs, fixtures, and templates.

2.0 APPLICATION

Applies to all tools used to produce deliverable goods.

3.0 ASSOCIATED MATERIALS

3.1   Limited Use Tooling Tag, Form No. B-43

3.2   Tool Inspection Report, Form No. B-44

4.0 PROCEDURE

4.1   All production tools such as jigs, fixtures, and templates used for producing deliverable goods must be inspected prior to use.

4.2   Tools that are also to be used for inspection purposes must also be calibrated by the measuring equipment control department (see Quality Procedure 4.3.1).

4.3   Tools are to be inspected in terms of their tool designs and work orders.

4.4   Unless otherwise specified, tools are to be proofed. For this purpose, they are to be tagged with a Limited Use Tooling Tag (Figure B-43), which (1) shows the tool number, (2) describes the limited purpose for which the tool can be used, and (3) displays the inspector's stamp or initial and date.

4.5   Parts produced as a result of proofing are to be inspected using the appropriate engineering drawings and/or specifications.

4.6   An accepted tool is to be identified as such with the inspector's stamp mark or initials on the tool nameplate. The Limited Use Tooling Tag is to be removed at this point.

4.7   If a tool is rejected, the inspector is to fill out a Tool Inspection Report and route the tool for rework. The form for this report, shown in Figure B-44, is to be filled out as follows: (1) the report number, (2) the tool number, (3) the tool name, (4) the date, (5) the number of the part for which the tool is used, (6) the

SUBJECT: TOOLING INSPECTION		
Authorized By: *V. A. Philis*	**Number:** 6.2.5	
Effective Date: 2/10/97	**Revision:** 1/30/97	**Page** 2 **of** 3

part name, (7) the work order used to manufacture the tool, (8) the name of the department or vendor that manufactured the tool, (9) the name of the inspector, (10) the reason for the rejection, and (11) rework instructions. At this point, the Limited Use Tooling Tag is to be replaced with a Rejection and Disposition Tag (Form No. B-49).

SUBJECT: TOOLING INSPECTION		
Authorized By: *V. A. Philis*		Number: 6.2.5
Effective Date: 2/10/97	Revision: 1/30/97	Page _3_ of _3_

Figure B-43. Limited Use Tooling Tag

Tool No. (1)_____

This tool can be used for the following limited purpose only: (2)

_____

_____

_____

Date _____

Inspection stamp or authorized signature (3)

Figure B-44. Tool Inspection Report

Report No. (1)	Tool No. (2)	Tool name (3)	Date (4)
Used for part No. (5)	Part name (6)	Work order No. (7)	Inspected by (9)
Manufactured by (department or supplier) (8)			
Reason for rejection (10)			
Rework instructions (11)			

SUBJECT: FIRST ARTICLE INSPECTION—IN-HOUSE MATERIAL		
Authorized By: *V. A. Philis*	**Number:** 6.3.1	
Effective Date: 2/10/97	**Revision:** 1/30/97	**Page** __1__ **of** __2__

1.0 PURPOSE

To define a method that the XYZ company uses for the inspection of the first of a large quantity of articles to be produced, the purpose being the early detection and timely correction of discrepancies.

2.0 APPLICATION

Applies to the first of all new or significantly changed deliverable parts or assemblies manufactured (produced) by the company. The first article inspection of subcontractor-produced items is covered by Quality Procedure 6.1.2.

3.0 PROCEDURE

3.1   First article inspection will normally be performed in the in-process inspection area for parts and assemblies that require additional work or assembly steps or in the final inspection area for finished goods.

3.2   The first article inspection is to be complete, covering all appropriate physical and functional characteristics of the part or assembly, workmanship, and the completeness and correctness of the required documentation.

3.3   Inspection will be accomplished in accordance with approved engineering drawings and specifications and, if appropriate, in accordance with special instructions from the quality plans and procedures department.

3.4   At the completion of the first article inspection, the inspector is to prepare the First Article Report—In-House Material. A copy of the form is shown in Figure B-45. It is to be filled out as follows: (1) the part or assembly name; (2) the date of the report; (3) the report number; (4) the part or assembly number; (5) the project or program name; (6) the name of the person who has submitted the article for inspection; (7) that person's department; a check mark either at (8) indicating approval or at (9) indicating rejection of the part or assembly; and, if rejected, (10) a description of the discrepancy; (11) a statement as to the recommended disposition; (12) a check mark to indicate whether or not a test report is attached; (13) the dated signature of the inspector; and (14) the dated signature of the person approving the report.

SUBJECT: FIRST ARTICLE INSPECTION—IN-HOUSE MATERIAL		
Authorized By: *V. A. Philis*		**Number:** 6.3.1
Effective Date: 2/10/97	**Revision:** 1/30/97	**Page** 2 **of** 2

Figure B-45. First Article Report—In-House Material

Part or assembly name (1)	Date (2)	Report No. (3)
Part or assembly No. (4)		Program/project name (5)
Submitted for inspection by (6)		Department (7)

First article approved (8) _____ This is your authority to proceed.

First article rejected (9) _____ Correct nonconformances and resubmit article.

Discrepancy (10)

Recommended disposition (11)

Detailed test report attached (12) _____ Yes _____ No

Inspected by (13)   Date   Approved by (14)   Date

SUBJECT: FINAL INSPECTION		
Authorized By: *V. A. Philis*		**Number:** 6.3.2
Effective Date: 2/10/97	**Revision:** 1/30/97	**Page** 1 **of** 2

1.0 PURPOSE

The procedure establishes the methods and responsibilities that the XYZ company uses for the final inspection of completed products to ensure that they comply with company standards and customer requirements.

2.0 APPLICATION

Applies to the inspection of all deliverable finished goods.

3.0 ASSOCIATED MATERIALS

3.1   Accepted Tag, Form B-46

3.2   Rejection and Disposition Tag, Form B-49

4.0 PROCEDURE

4.1   All finished goods are to be presented to the final inspection department for inspection. These inspections are to take place when it is determined that a special inspection is necessary.

4.2   Parts and assemblies will not be accepted for final inspection unless all operations called out on the operations instructions traveler are identified as complete.

4.3   Inspections are to be made using applicable inspection instructions, drawings, specifications, and other appropriate reference materials.

4.4   The inspection is to include an examination of the accompanying paperwork for completeness and correctness, workmanship, physical and functional characteristics, and the proper markings on parts and assemblies.

4.5   When specified, parts or assemblies found to be acceptable are to be so marked using either an inspection steel stamp, an inspection rubber stamp, electric etching, or an Accepted Tag (Form No. B-46). This information is also to be recorded on the accompanying instructions.

4.6   When the Accepted Tag, Figure B-46, is necessary, it is filled out as follows: (1) part/detail number, (2) date, (3) part or detail name, (4) department or vendor, (5) work order or purchase order, (6) rejected, (7) accepted, (8) remarks, and (9) inspection stamp or initials of the person issuing the tag.

4.7   Materials to be reworked are to be routed to the appropriate department together with rework instructions.

SUBJECT: FINAL INSPECTION			
Authorized By: *V. A. Philis*		**Number:** 6.3.2	
Effective Date: 2/10/97	**Revision:** 1/30/97	**Page** 2 **of** 2	

Figure B-46. Accepted Tag

Part No. (1)	Date (2)
Part name (3)	
Department or supplier (4)	
Work order or P.O. (5)	
Quantity rejected (6)	Accepted (7)
Remarks (8)	
Authorized signature or stamp (9)	

SUBJECT: FINAL ACCEPTANCE TESTING			
Authorized By: *V. A. Philis*		**Number:** 6.3.3	
Effective Date: 2/10/97	**Revision:** 1/30/97	**Page** __1__ **of** __1__	

1.0 PURPOSE

This procedure describes the methods and responsibilities that the XYZ company uses for conducting the final acceptance testing of finished goods to ensure that they comform to the XYZ company's and the customer's standards.

2.0 APPLICATION

Applies to the formal acceptance testing of all finished goods when testing has been specified.

3.0 ASSOCIATED MATERIAL

3.1   Final Acceptance Processing, Quality Procedure 6.3.4

4.0 PROCEDURE

4.1   Products are to be approved for acceptance testing by final inspection and test after a determination has been made that the product is complete, that all operations have been completed, and that there are no open items.

4.2   Each acceptance test is to be witnessed by the customer or the customer's representative unless otherwise specified.

4.3   Final inspection and test is responsible for monitoring the test area to make sure test equipment that has passed its "calibration due" date is not used.

4.4   Completed products requiring testing are to be tested in accordance with customer's specifications. This is usually a 20-hour run.

4.5   Failures are to be recorded on the Failure Report, Form No. B-42 (see Quality Procedure 6.2.4).

4.6   Remove the details that are either the cause of or damaged by the failure.

4.7   Materials to be reworked are to be routed to the appropriate department together with rework instructions.

SUBJECT: FINAL ACCEPTANCE PROCESSING		
Authorized By: *V. A. Philis*		**Number:** 6.3.4
Effective Date: 2/10/97	**Revision:** 1/30/97	**Page** __1__ **of** __3__

1.0 PURPOSE

To set forth the requirements that the XYZ company uses for the final acceptance processing of completed goods.

2.0 APPLICATION

Applies to the processing of all completed goods prior to shipment.

3.0 ASSOCIATED MATERIALS

3.1   Acceptance, Comments, and Conditions, Form No. B-47

3.2   Deliverable Documentation Review, Quality Procedure 4.1.3

4.0 PROCEDURE

4.1   After successful completion of final inspection and test, completed products are to be examined for the following:

4.1.1   Completeness and correctness of documentation (see Quality Procedure 4.1.3).

4.1.2   The presence, when specified, of the proper inspection stamps.

4.1.3   The cleanliness of the finished product.

4.1.4   The quality of the finished surfaces.

4.1.5   The absence or presence of handling damage.

4.1.6   Missing parts.

4.1.7   The presence and correctness of nameplates, when specified.

4.1.8   The presence and correctness of instruction plates, when specified.

4.1.9   The presence of safety guards. In the event that guards are not supplied by the XYZ company, however, the inspector must secure a release from appropriate sources.

SUBJECT: FINAL ACCEPTANCE PROCESSING		
Authorized By: *V. A. Philis*	**Number:** 6.3.4	
Effective Date: 2/10/97 **Revision:** 1/30/97	**Page** _2_ **of** _3_	

4.2   For a product that must comply to a weight requirement, the inspector is to weigh the item and record the amount when specified.

4.3   The inspector is to resolve any and all open items.

4.4   The inspector is to see that, if spare parts are to be shipped with the product, they are located and transported to the shipping area. The inspector is also to make sure these spares have been inspected.

4.5   If special packaging is required, the inspector is to witness this process to make sure specifications are in compliance.

4.6   When satisfied that the finished goods are in conformance with requirements, the inspector is to affix his or her initials to the appropriate paperwork.

4.7   When specified, an Acceptance, Comments, and Conditions Form is to be filled out by the inspector and signed by the customer or the customer's representative. An example of the form used for this purpose is shown in Figure B-47. It is to be filled out as follows: (1) the part number, (2) the part name, (3) the serial number, (4) the program or project, (5) the contract, (6) the acceptance date, (7) a check mark to indicate that the product is being accepted "as is" or "with conditions," (8) comments, (9) conditions, (10) person accepting the product, (11) that person's title, (12) the name of the organization or firm that person is representing, (13) acceptance of acknowledger's signature, (14) title of accepter, (15) department of accepter, (16) a description of follow-up, (17) the name of the person responsible for the follow-up, (18) title of that person, and (19) department of that person. A copy of this completed form is to be sent to the finance and administration department.

SUBJECT: FINAL ACCEPTANCE PROCESSING		
Authorized By: *V. A. Philis*		**Number:** 6.3.4
Effective Date: 2/10/97	**Revision:** 1/30/97	**Page** _3_ **of** _3_

Figure B-47. Acceptance, Comments, and Conditions

Part No. (1)	Part name (2)	Serial No. (3)
Program/project (4)	Contract (5)	Acceptance date (6)

Accepted (7)   _____ *as is*   _____ *with conditions*

Comments (8)

Conditions (9)

Accepted (10) Date	Title (11)	Representing (12)
Acknowledged by (13) Date	Title (14)	Department (15)

Follow-up (16)

Follow-up by (17) Date	Title (18)	Department (19)

SUBJECT: PROTECTING PRODUCT QUALITY			
Authorized By: *V. A. Philis*		**Number:** 6.4.1	
Effective Date: 2/10/97	**Revision:** 1/30/97	**Page** 1 **of** 2	

1.0 PURPOSE

To describe a system that the XYZ company uses for specifying and monitoring proper storage, packaging, and shipping practices to protect the quality of deliverable products and to prevent their damage, deterioration, and degradation.

2.0 APPLICATION

Applies to all deliverable items, including spares.

3.0 ASSOCIATED MATERIALS

3.1   Quality Planning, QP 3.2.2

3.2   Inspection/Test Instructions, QP 3.3.1

4.0 PROCEDURE

4.1   The quality department and/or project manager is responsible for specifying, when required, instructions for the proper handling, preservation, storage, packaging, and shipping of goods to protect quality and prevent damage (see QP 3.2.2).

4.2   The quality department is responsible for seeing that the quality plan specifications are related to protecting product quality and are reflected in the detailed inspection and test instructions (see QP 3.3.1).

4.3   The project manager is responsible for monitoring the handling of goods throughout the manufacturing process to ensure that both specified and common sense practices are being followed. For instance, heavy or bulky items should be mounted on proper skids, delicate items should be prevented from jostling damage, and dangerous stacking of in-process materials must be prohibited.

4.4   The project manager is responsible for periodically, and randomly, auditing the storage of goods to determine if goods are being stored safely, if specified environmental conditions are being maintained, if goods are being properly rotated, if obsolete and out-of-date goods are being removed, and if goods are properly identifed.

4.5   The project manager is responsible for overseeing the proper packaging of finished goods prior to shipment. This includes making sure that packaging specifications are in compliance, including the use of dehydrating agents and humidity indicators, if required, and making certain that all interior and exterior containers are properly marked.

SUBJECT: PROTECTING PRODUCT QUALITY		
Authorized By: *V. A. Philis*		**Number:** 6.4.1
Effective Date: 2/10/97	**Revision:** 1/30/97	**Page** _2_ **of** _2_

4.6   The project manager is responsible for overseeing the proper shipping of goods, including verifying that the shipping documents are correct and properly stamped, that containers are properly loaded and secured in shipping vehicles, and that there is compliance with Interstate Commerce Commission regulations and other applicable regulations.

SUBJECT: MATERIAL REVIEW BOARD			
Authorized By: *V. A. Philis*		**Number:** 6.5.1	
Effective Date: 2/10/97	**Revision:** 1/30/97	**Page** 1 **of** 1	

1.0 PURPOSE

To set forth the organization and responsibilities of a material review board (MRB) at the XYZ company.

2.0 APPLICATION

Applies to the engineering, manufacturing (production), material, and quality assurance departments as related to the processing of materials that have been rejected by the inspection or test functions as nonconforming to specified requirements.

3.0 PROCEDURE

3.1   Members of the material review board will be selected from the quality assurance department and the engineering department. In appropriate instances, customer representatives may also be requested to participate.

3.2   The primary responsibility of the material review board is to decide on the disposition of nonconforming materials.

3.3   When appropriate, board members may call on the services of other company employees to serve on the MRB in a nonvoting, advisory capacity.

3.4   For a contracted item, the MRB's disposition decision must receive the concurrence of the customer.

3.5   The quality assurance function is responsible for providing the MRB with a material review area for use in holding and accounting for materials that are awaiting MRB action.

SUBJECT: MATERIAL REVIEW BOARD DOCUMENTATION		
Authorized By: _V. A. Philis_	**Number:** 6.5.2	
Effective Date: 2/10/97	**Revision:** 1/30/97	**Page** 1 **of** 1

1.0 PURPOSE

To provide a single list and description of use of the documents used in the material review board system at the XYZ company.

2.0 APPLICATION

Applies to all activities related to the rejection of nonconforming materials, their review, corrective action, and disposition.

3.0 ASSOCIATED MATERIALS

3.1   Rejection and Disposition Tag, QP 6.5.3, Form No. B-49

3.2   Material Review Process, QP 6.5.3, Form No. B-48

3.3   Request for Deviation or Waiver, QP 6.5.4, Form No. B-50

4.0 PROCEDURE

4.1   Material received for material review board (MRB) action arrives with Rejection and Disposition Tag (Form No. B-49).

4.2   After initial examination of material, a Material Review Report (MRR) is initiated (QP 6.5.3).

4.3   The MRB decides on the material's disposition and, when appropriate, secures customer concurrence, with all of this recorded on the previously mentioned Material Review Report (QP 6.5.3).

4.4   Materials are transmitted to their appropriate locations according to the decided disposition (QP 6.5.3). Materials to be scrapped are processed (QP 6.5.6).

4.5   When applicable, Requests for Deviations or Waivers are processed (QP 6.5.4).

4.6   When applicable, requests for Authorization to Use Equivalent Items are processed (QP 6.5.5).

SUBJECT: MATERIAL REVIEW PROCESS		
Authorized By: *V. A. Philis*		**Number:** 6.5.3
Effective Date: 2/10/97	**Revision:** 1/30/97	**Page** __1__ **of** __4__

1.0 PURPOSE

This procedure describes the system for reviewing materials that have been found to be nonconforming and to order appropriate dispositions of those materials.

3.0 DEFINITIONS

3.1   Use-as-is material: Material that is found to be nonconforming in a minor way, but suitable for its intended purpose and acceptable to the customer.

3.2   Reworkable material: Nonconforming material that can be reworked to an acceptable condition.

3.3   Scrap material: Nonconforming material that cannot be reworked to an acceptable condition.

3.4   Deviation: Permission in writing to deviate from specified requirements (see Request for Deviation or Waiver, QP 6.5.4).

3.5   Waiver: Permission in writing to accept for use a completed but nonconforming item either "as is" or on completion of rework (see Request for Deviation or Waiver, QP 6.5.4).

3.6   Equivalent item: An item that is interchangeable with and equal to or better than the item called for in the specification (see Authorization to Use Equivalent Item, QP 6.5.6).

4.0 ASSOCIATED MATERIALS

4.1   Rejection and Disposition Tag, Form No. B-49

4.2   Material Review Report (MRR), Form No. B-48

5.0 PROCEDURE

5.1   Material received for action by the material review board (MRB) is to be first examined by the board's quality assurance representative to determine if the material and accompanying documentation are in order and that the material has been properly tagged with a Rejection and Disposition Tag (see Paragraph 5.8 of this section). If the material or documentation is not in order, the quality assurance representative is responsible for taking corrective action.

SUBJECT: MATERIAL REVIEW PROCESS		
Authorized By: *V. A. Philis*		**Number:** 6.5.3
Effective Date: 2/10/97	**Revision:** 1/30/97	**Page** 2 **of** 4

5.2   The MRB's quality assurance representative is responsible for the initiation of the Material Review Report Form, illustrated in Figure B-48, filling in parts 1 through 15: (1) Material Review Report number and page number; (2) the part number of the discrepant part; (3) the part's name; (4) the program name or project name; (5) the contract number or work order number; (6) the department where the discrepancy was detected or, in the case of purchased items, the name and purchase order number of the supplier; (7) the quantity of pieces rejected; (8) affected serial numbers, if applicable; (9) the current location of the items; (10) the part's specification number and revision number; (11) the part's drawing number and revision number; (12) an indication of whether or not there is restricted end use or (13) whether or not the next assembly is affected; and (14) a description of the nonconformance together with an explanation of the probable cause; and (15) the authorized disposition.

5.3   When required, the MRB's quality assurance representative is responsible for securing the concurrence of quality assurance and/or engineering and/or the customer to the selected disposition. Concurrence is indicated by signature and date (space 16) as appropriate and applicable.

5.4   When required, mark original Rejection and Disposition Tag (see Paragraph 5.1) with appropriate Material Review Acceptance Disposition.

5.5   Materials and documentation are now to be transmitted to the appropriate locations for disposition.

5.6   When the approval is appropriately signed, item (17) calls for the appropriate CAR information, who wrote the requisition, the initiating department and the date.

5.6   After an appropriate interval of time, to be determined separately for each case by the MRB, verification of the authorized disposition should be made, with the verifier recording findings and signing and dating this in item (18) of the MRR.

5.8   In case the rejection and disposition tag is needed (Figure B-49), it is to be filled in as follows: (1) part number, (2) date, (3) part name, (4) department or vendor, (5) work order or purchase order, (6) rejected, (7) accepted, (8) reason, and (9) inspection stamp or initials of the person issuing the tag. Reflect verification of disposition by recording the name of the verifier (10) and the date when the verification was made (11).

SUBJECT: MATERIAL REVIEW PROCESS		
Authorized By: *V. A. Philis*		**Number:** 6.5.3
Effective Date: 2/10/97	**Revision:** 1/30/97	**Page** 3 **of** 4

Figure B-48. Material Review Report

Number (1)	Part No. (2)	Part name (3)
Page ____of	Program/project (4)	

Contract or work order (5)	Department or supplier and purchase order (6)	
Quantity of pieces (7)	Serial No. (8)	Item location (9)

Specification and revision (10)	Drawing No. and revision (11)

Restricted end use (12) _____ Yes _____ No	Next assembly affected (13) _____ Yes _____ No

Description of nonconformance and apparent cause (14)

Authorized disposition (15)

Material review approval (16)

Quality assurance ____ Date
Engineering ____ Date
Customer ____ Date

CAR No.   CAR initiated   Written by (17)   Department   Date

Disposition verification (18)

_____
Verifier   Title   Date

SUBJECT: MATERIAL REVIEW PROCESS		
Authorized By: *V. A. Philis*	**Number:** 6.5.3	
Effective Date: 2/10/97	**Revision:** 1/30/97	**Page** 4 **of** 4

Figure B-49. Rejection and Disposition Tag

Part No. (1)	Date (2)
Part name (3)	
Department or supplier (4)	
Work order or purchase order (5)	
Quantity rejected (6)	Accepted (7)
Remarks (8)	
Authorized signature or stamp (9)	

SUBJECT: REQUEST FOR DEVIATION OR WAIVER			
Authorized By: *V. A. Philis*		**Number:** 6.5.4	
Effective Date: 2/10/97	**Revision:** 1/30/97	**Page** 1 **of** 3	

1.0 PURPOSE

To provide instructions for the XYZ company to request from a customer advance permission to deviate from specified requirements or a waiver of specified requirements for an already manufactured detail or assembly.

2.0 APPLICATION

Applies to customers and to quality assurance, engineering, manufacturing (production), the material review board, and marketing's contracts department.

3.0 DEFINITIONS

3.1   Deviation permission: Permission, in writing and in advance of manufacture (production), to deviate from specified requirements.

3.2   Waiver: Permission, in writing, to accept for use a completed, but nonconforming, item either "as is" or on completion of rework.

3.3   Minor, major, critical: Deviations and waivers are classified as minor, major, or critical as defined by mutual consent between the XYZ company and the customer.

4.0 ASSOCIATED MATERIALS

4.1   Request for Deviation or Waiver, Form No. B-50

5.0 PROCEDURE

5.1   The material review board prepares the Request for Deviation or Waiver Form (Figure B-50) as follows: (1) the number of the request; (2) the date the request is directed; (3) the name, address, and telephone number of the requesting function (customer); (4) who is sending this request; (5) the type of request; (6) the criticality; (7) the name of the program or project; (8) the contract number; (9) the part number; (10) the part name; (11) the drawing or specification number; (12) the affected units(s) or dates of the deviation or waiver; (13) the explanation of the reason for the deviation or waiver; and (14) the appropriate requesting signatures.

SUBJECT: REQUEST FOR DEVIATION OR WAIVER		
Authorized By: *V. A. Philis*		**Number:** 6.5.4
Effective Date: 2/10/97	**Revision:** 1/30/97	**Page** __2__ **of** __3__

   5.2   The completed form is next presented to the customer for approval. Approval is indicated by (15) an authorized customer signature, title, and date on the Request for Deviation or Waiver.

   5.3   On the MRB's receipt of the returned request, approved (signed) or rejected (unsigned), it is to be forwarded to the appropriate department for disposition and to be made part of the job file.

SUBJECT: REQUEST FOR DEVIATION OR WAIVER		
Authorized By: *V. A. Philis*		**Number:** 6.5.4
Effective Date: 2/10/97	**Revision:** 1/30/97	**Page** _3_ **of** _3_

Figure B-50. Request for Deviation or Waiver

Number (1)	Date (2)
To (customer) (3)	From (4)
Type of request (5) ___ Deviation ___ Waiver	Criticality (6) ___ Minor _____ Major ____ Critical
Program/project (7)	Contract or work order (8)

Part No. (9)	Part name (10)	Drawing or specification (11)

Affects serialized assembly number (12) _____ Only
or From serialized assembly number _____ through _____
or Production dates _____ through _____

Explanation (13)

Requested by (14)
 	
    Engineering   Date
 	
    Quality assurance Date
	Approved by customer (15)
	
	Name _____
	
	Title _____ Date_____

SUBJECT: AUTHORIZATION TO USE EQUIVALENT ITEM		
Authorized By: *V. A. Philis*		**Number:** 6.5.5
Effective Date: 2/10/97	**Revision:** 1/30/97	**Page** __1__ **of** __2__

1.0 PURPOSE

To provide a system for the XYZ company for approving, on an emergency basis, the limited use of a substitute equivalent item in lieu of the specified item.

2.0 APPLICATION

This procedure applies only to the temporary substitution of an item.

3.0 DEFINITIONS

3.1   Equivalent item: A completely interchangeable item that is equal to or better than what is specified.

4.0 ASSOCIATED MATERIAL

4.1   Equivalent Item Authorization, Form No. B-51

5.0 PROCEDURE

5.1   The Equivalent Item Authorization must be processed prior to the use of a substitute item.

5.2   The Equivalent Item Authorization (EIA) Form (Figure B-51) is filled out as follows, normally by a representative of the manufacturing (production) department: (1) the EIA number, (2) the name of the program or project, (3) the drawing and/or specification number, (4) the identification by number and name of the specified part and the recommended substitute part, (5) the limitation as to the assembly or assemblies for which the substitution would be permissible, (6) a description of the purpose of the substitution, and (7) the signature of the requester and the date of the request.

5.3   The Equivalent Item Authorization Form is limited to the listing of a single part number.

5.4   The requester is required to secure the approval signatures of appropriate representatives from (8) quality assurance, (9) engineering, and (10) manufacturing (production).

SUBJECT: AUTHORIZATION TO USE EQUIVALENT ITEM		
Authorized By: *V. A. Philis*		**Number:** 6.5.5
Effective Date: 2/10/97	**Revision:** 1/30/97	**Page** 2 **of** 2

Figure B-51. Equivalent Item Authorization

Number (1)	Program/project (2)
Drawing/specification No. (3)	

Change part No. (4) _____ Name _____
To part No. _____ Name _____

Effective for serialized assembly number (5) _____ Only
or From serialized assembly number _____ through _____

Purpose of substitution (6)

Requested by (7) Date	Approved quality assurance (8) Date
Approved engineering (9) Date	Approved production (10) Date

SUBJECT: SCRAP CONTROL		
Authorized By: *V. A. Philis*		**Number:** 6.5.6
Effective Date: 2/10/97	**Revision:** 1/30/97	**Page** _1_ **of** _1_

1.0 PURPOSE

To describe the methods that the XYZ company uses for processing scrap materials.

2.0 APPLICATION

Applies to the salvage unit of the quality management and administration function. Applicable to all materials designated as scrap, except customer-owned scrap material.

3.0 DEFINITIONS

3.1   Scrap: Material that has been found to be unfit for further processing.

4.0 ASSOCIATED MATERIALS

4.1   Material Review Process, QP 6.5.3

5.0 PROCEDURE

5.1   Material that has been clearly marked, stamped, and/or tagged as scrap is to be delivered to the scrap material area.

5.2   Scrap material personnel are responsible for verifying that the received material has been correctly and completely identified as scrap.

5.3   In cases for which vendor responsibility has been established for the scrapped condition of vendor-supplied materials, those materials are to be returned to the vendor for credit or replacement.

5.4   The material is to be analyzed to determine if all or part of it can be salvaged for other uses. Consideration should be given to such alternate uses of scrapped fabricated parts or assemblies for display purposes, instructional purposes, test purposes, and machinery setup purposes.

5.5   Any item salvaged for alternate use purposes must be clearly and prominently identified so as not to confuse it with an accepted item (mutilation is a viable option).

5.6   Materials for which no practical alternative uses can be found are to be disposed of in an efficient, cost-effective manner.

5.7   The salvage unit of the quality management and administration function is to operate a system to publicize to employees the policy of scrap prevention and conservation of resources.

SUBJECT: STATISTICAL AND RELIABILITY QUALITY CONTROL		
Authorized By: *V. A. Philis*	**Number:** 6.6.1	
Effective Date: 2/10/97	**Revision:** 1/30/97	**Page** 1 **of** 2

1.0 PURPOSE

The purpose is to describe the various statistical tools used by quality assurance.

2.0 APPLICATION

Applies to all parts produced.

3.0 DEFINITIONS

3.1   Machine capability: The study of short-term variation of a given characteristic. It is a measurement of the amount of random variation in a machine and excludes all sources of assignable variation. Typically, it consists of 125 data points collected consecutively. It provides an optimistic estimate of the variation in a characteristic given the tightest control possible.

3.2   Process capability: The study of long-term variation in a given characteristic. Typically, it consists of 125 data points collected at intervals spanning several days or weeks. The results from a process capability study provide a realistic estimate of the variation in a characteristic given normal day-to-day controls.

3.3   Sampling inspection: A determination of the probable quality of all items in a single group on the basis of examining the quality characteristics of a selected segment of that group.

4.0   PROCEDURE

4.1   The quality control department is responsible for generating and evaluating information to determine if a process is capable of manufacturing quality parts. Quality control will also determine if any type of process controls are necessary to maintain capability, reliability, and maintainability. Types of process controls may include, but are not limited to, machine capability studies, X and R charts, percentage defective charts, median charts, process capability studies, MTBF, reliability studies, and durability life studies.

4.2   The quality control department is responsible for overseeing that proper training and instructions are given for the use of these statistical tools.

4.3   When choosing process controls, quality control shall consider customer requirements, machine capability, the nature of the machining operation, and any past problems in producing the part.

4.4   Inspection by statistical sampling can be applied, as appropriate, and, when specified, in receiving inspection, in-process inspection, and final inspection.

SUBJECT: STATISTICAL AND RELIABILITY QUALITY CONTROL		
Authorized By: *V. A. Philis*		**Number:** 6.6.1
Effective Date: 2/10/97	**Revision:** 1/30/97	**Page** _2_ **of** _2_

4.5   Only sampling plans approved by the quality control department can be used.

4.6   Approved sampling plans, unless otherwise stated, will conform to the XYZ company requirements and those of their customers.

4.7   When applying a sampling plan, inspectors are to select samples carefully and randomly from a given lot.

4.8   The statistical sampling plan can be tightened or loosened temporarily or permanently when an examination of quality control records indicates that such action is warranted.

4.9   Approval of R&M studies are suggested and controlled by the quality assurance department (see also Procedure 7.1.1).

SUBJECT: QUALITY RELIABILITY REPORTING PROCESS			
Authorized By: *V. A. Philis*		**Number:** 7.1.1	
Effective Date: 2/10/97	**Revision:** 1/30/97	**Page** 1 **of** 1	

1.0 PURPOSE

To provide a consistent quality and reliability progress reporting format at XYZ company.

2.0 APPLICATION

This procedure applies to all facilities and departments at XYZ company. The intent of this procedure is to make sure customer requirements are satisfied.

3.0 PROCEDURE

3.1 Data sources: Some of the appropriate and common data sources to utilize for quality and reliability data are

MTBF	mean time between failures
MCTF	mean cycles to failure
MTTR	mean time to replace (or repair)
RATS	returns analysis tracking system
PTTR	personalized time to repair

3.2 Format: The charts maintained by customer satisfaction include the following criteria description:

Performance: The cycle of performance is evaluated at 90, 180, and 360 days after start or as customer specified.

Actual/projected forecast: The actual or projected forecast figure depends on contractual agreement between XYZ and the customer.

%: Percentage of warranty claims is based on analysis of actual data.

Analysis: Based on information received after analyzing the product or the system.

Corrective action: This section includes the efforts of the XYZ company to identify, correct, and provide a prevention approach to the root cause problems.

Forecast improvement: An individual improvement forecast for a given project must total the same number used in the actual/projected forecast.

SUBJECT: RECORD RETENTION			
Authorized By: *V. A. Philis*		**Number:** 7.1.2	
Effective Date: 2/10/97	**Revision:** 1/30/97	**Page** 1 **of** 2	

1.0 PURPOSE

This procedure defines the record retention requirements for documents related to the XYZ total quality system.

2.0 APPLICATION

This procedure applies to the entire corporation.

3.0 PROCEDURE

3.1 The record index attached in this procedure will define those documents to be retained and the length of time to be retained. The following guidelines are to be used in conjunction with this record index:

　　　3.1.1 Unless specified otherwise, records created during a particular project are grouped together and retained for the time specified.

　　　3.1.2 When more than one date appears on a record, the retention period is based on the latest date on that record.

　　　3.1.3 Obsolete procedures and written instructions are those that have been replaced by a new version. The retention period applies to the historical file of obsolete procedures and instructions.

3.2 Each record is to be destroyed as soon as practicable following the specified retention period. However, when longer retention periods are required due to patent, legal, or other corporate activities, as well as contractual agreements, these are to be given preference over this schedule.

3.3 Records may be retained on film, electronically, and/or on hard copy as required on a project-by-project basis. Records are to be stored safely so that they may be readily accessible and retrievable for use in a reasonable amount of time. Storage areas for retained records should provide adequate protection from unauthorized access, moisture, and fire.

SUBJECT: RECORD RETENTION		
Authorized By: *V. A. Philis*	**Number:** 7.1.2	
Effective Date: 2/10/97 **Revision:** 1/30/97	**Page** 2 **of** 2	

Record Index

Record description	Retention beyond current project (years)
Receiving inspection records	1
Plant laboratory test/inspection records	1
Source and subcontractor surveillance records	2
Supplier correspondence of nonconforming material	0.5
Initial sample approval records	While project is active + 1
Gage control records	2
Procedures and approvals (master)	5
Certification of conformance to governmental regulations and/or standards	
From suppliers	1
To customer	1
Material review/variation authority	5
Process capability and process performance studies	1
Reliability studies	1
Written work/inspection instructions	1 year after revision
In-process and final inspection and/or test records	0.5
Corrective action reports traceable to inspection records	1
Final disposition of material traceable to inspection records	1
Customer, supplier, and internal quality systems audit reports, work papers, and corrective action replies	1

SUBJECT: CHARACTERISTIC INSPECTION		
Authorized By: *V. A. Philis*	**Number:** 7.1.3	
Effective Date: 2/10/97	**Revision:** 1/30/97	**Page** 1 **of** 2

1.0 PURPOSE

Define the procedure for characteristic inspection and reporting of carbide/ceramic, PCD, CBN, cutting tools as applicable to all carbide/ceramic, PCD, CBN, cutting tools.

2.0 RELATED DOCUMENTS

2.1 Appropriate blueprint

2.2 Final inspection report (INSP XXX)

3.0 DEFINITIONS

3.1 Critical characteristic: A characteristic directly responsible for dimensional accuracy of the piece/part being produced.

3.2 Functional characteristic: A characteristic not directly affecting the accuracy or performance of the piece/part.

4.0. PROCEDURE

4.1 Inspect critical characteristics by geometry
        4.1.1 Triangular
                a. Base to apex
                b. Inscribe circle
        4.1.2 Square, hexagon, pentagon, octagon
                a. Inscribe circle
                b. Gage pin
        4.1.3 Diamond
                a. Inscribe circle
                b. Gage pin
                c. Radius to radius (optional)
        4.1.4 Round
                a. Inscribe circle
        4.1.5 Rectangular
                a. Width
                b. Length
4.2 Threaders and groovers will have those critical characteristics inspected that control the form geometry (e.g., width on groover or included angle, root truncation on threader).

SUBJECT: CHARACTERISTIC INSPECTION		
Authorized By: *V. A. Philis*		**Number:** 7.1.3
Effective Date: 2/10/97	**Revision:** 1/30/97	**Page** 2 **of** 2

4.3 Cutting tools designated to generate a form shall have the appropriate form control characteristic inspected.

4.4 Clearance angles shall be considered a functional characteristic.

4.5 Unless otherwise stated, all critical characteristics will be maintained by final grind (see Procedure XY). Any subsequent operation to grind (e.g., hone or coating application) could affect dimensional accuracy.

4.6 When critical characteristics are interrelated through the method of manufacture, only one of the characteristics needs to be inspected (e.g., for cam ground triangle, certify inscribed circle or base to apex; the other characteristics will be considered functional).

4.7 Characteristic inspection technique.
> 4.7.1 SPC shall be used whenever possible.

> 4.7.2 Dimensional inspection shall be accomplished through the use of basic inspection equipment or SPC data-gathering devices if applicable.

> 4.7.3 Dimensional inspection will be accomplished in accordance with specific departmental procedures.

4.8 Documentation
> 4.8.1 Inspection data will be transported from the in-process sheets or tapes onto the Valentine Inspection Form (INSP ZZZ).

> 4.8.2 Only the average and range for each characteristic will be recorded on the inspection form.

> 4.8.3 Distribution of copies will be as follows:
> Customer (white)   Attach to packing list
> Quality assurance (yellow)   Send to quality assurance department
> Manufacturing (tan)   Retain at manufacturing location

Glossary

In this glossary, we provide the reader with some of the most common words used in the quality world of the tooling and equipment industry. By no means is this list exhaustive. Rather, it is a list that will facilitate the understanding of a quality system in the TE industry. For an official explanation of the words, the reader is encouraged to see the *Reliability and Maintainability Guidelines* and ANSI/ASQC A8402:1994 or ISO 8402:1994.

Acceptance test. A test to determine machinery/equipment conformance to the qualification requirements in its equipment specifications.

Appraisal costs. The costs of assessing the quality achieved.

Assessment. The evaluation of a company's management system against the requirements of a set of criteria to evaluate its level of compliance.

Audit. A systematic and independent examination to determine whether quality activities and related results comply with planned arrangements and whether these arrangements are implemented effectively and are suitable to achieve objectives. *See also* Compliance audit, External audit, Extrinsic audit, Gap audit, Internal audit, Product audit, Systems audit

Audit performance against company objectives. A review based on the performance of activities programs or projects against specific company objectives. This may include a management review of quality performance against the objectives.

Audit of policies and objectives. This audit covers the business aspects of the company's quality and its technological aspects. Policies and objectives should be clearly stated in a quality manual or equivalent document.

Auditor. A person who is qualified to perform all or any part of a quality systems audit.

Benchmark audit. See Gap audit

Calibration. All the operations for the purpose of determining the values of the errors of a measuring instrument (and, if necessary, to determine other metrological properties).

Certification system. A system having its own rules for procedures and management for performing conformance certification.

Compliance. An indication or judgment that the product or service meets the requirements of the relevant specification or regulation; also the state of meeting the requirements.

Compliance audit. A review to see if the implementation of the quality system follows procedures, work instructions, quality plan, quality manual, standards, and so on.

Conceptual design: The design process in which concepts are generated with a view to fulfilling the objectives.

Conformance. An indication or judgment that the producer of service meets the requirements of a standard (such as QS-9000).

Design. To generate information from which a required product can become a reality (verb). The set of instructions (e.g., specifications, drawings, and schedules) necessary to construct a product (noun). For specified requirements, either of the following applies: (a) requirements prescribed by the purchaser and agreed by the supplier in a contract for products or services, or (b) requirements prescribed by the supplier that are perceived by the supplier as satisfying a market need.

Design evaluation. A systematic examination of the result of an activity to establish the degree to which the original objectives have been fulfilled.

Design review. A formal, documented, comprehensive, and systematic examination of a design to evaluate the design requirements and the capability of the design to meet these requirements and identify problems and propose solutions.

Detail design. The process in which the precise shape, dimension, and tolerance are specified, the material selection is confirmed, and the method of manufacture is considered for every individual part of the product.

Embodiment design. The design process in which a structured development of the preferred concept is performed. The preliminary embodiment of all the main functions to be performed by the product is undertaken, and the physical processes are clearly established.

External audit. An audit carried out by a company of its own suppliers or potential suppliers; also known as second-party audit.

Extrinsic audit. An audit performed by a customer or a regulatory body, such as a registration body, or an inspection agency; also known as a third-party audit.

Failure. An event that causes the machine/equipment not to meet its design intent.

Failure analysis. The logical, systematic examination of a failure.

Failure costs. Internal costs arise within an organization from the failure to achieve the quality specified; external costs arise outside an organization from the failure to achieve the quality specified.

First-party audit. See Internal audit

Gap audit (assessment). An audit to assess the gaps within a company's current quality system; also provides a baseline that can be used to compare improvements over specific periods of time.

Inspection. Activities such as measuring, examining, testing, and gaging one or more characteristics of a product or service and comparing these with specified requirements to determine conformance.

Instruction. The documented direction given in regard to what is to be done, including the information given in training.

Internal audit. Audit performed by a company on its own systems, procedures, and facilities; also known as a first-party audit.

ISO. Official title for the International Organization for Standardization. ISO is not an acronym; it is the Greek word *iso,* meaning *equal.*

Lead auditor. An auditor who has the qualification and is authorized to manage a quality system audit.

Maintainability. A characteristic of design, installation, and operation, usually expressed as the probability that a machine can be retained in, or restored to, the designed condition of operation within a specified interval of time when maintenance is performed according to prescribed procedures.

Management review. A formal evaluation by top management of the status and adequacy of the quality system in relation to quality policy and objectives resulting from changing circumstances.

Objective evidence. The qualitative or quantitative information, records, or statement of fact pertaining to the quality of an item or service or to the existence and implementation of a quality system element or documented requirement, based on observation, measurement, or test and which can be verified.

Prevention costs. The costs of any action taken to investigate, prevent, or reduce defects and failures.

Process. Any operation in a business.

Process capability. The limits of inherent variability within which a process operates as governed by the prevailing circumstances.

Product audit. An evaluation to determine if the product meets the specifications and needs of its use.

Product design. Output that an organization supplies to a user in the context of product design.

QS-9000 requirement. A quality system for the American automotive industry that is based on ISO 9001 with added industry- and customer-specific requirements.

Quality. The totality of features and characteristics of a product or service that bear on its ability to satisfy a stated or implied need.

Quality assurance. All the planned and systematic actions necessary to provide adequate confidence that a product or service will satisfy given requirements for quality.

Quality audit. See Audit

Quality control. The operational techniques and activities that are used to fulfill requirements for quality.

Quality evaluation. A systematic examination of the extent to which an entity is capable of fulfilling specified requirements.

Quality manual. A document stating the quality policy and describing the quality system of an organization.

Quality plan. A document setting out the specific quality practices, resources, and sequence of activities relevant to a particular product, service, contract, or project.

Quality policy. The overall intentions and direction of an organization with regard to quality as formally expressed by top management.

Quality-related costs. Costs incurred in ensuring and assuring satisfactory quality, as well as the losses incurred when satisfactory quality is not achieved. *See also* Appraisal costs, Prevention costs, Failure costs

Quality systems. The organization structure, responsibilities, procedures, processes, and resources needed to implement quality management.

Registration. The formal acknowledgment by a registration body accredited by a recognized national authority that an organization has been assessed and shown to comply with the ISO 9000 series of standards at the time of assessment.

Reliability growth. Machine reliability improvement as a result of identifying and eliminating machinery or equipment failure causes during machine testing and operation.

Sample. A group of items or individuals, taken from a larger collection or population, that provides information needed for assessing a characteristic (or characteristics) of a population or serves as a basis for action on the population or the process that produced it.

Second-party audit. See External audit

Supplier appraisal. Assessment of a supplier's capability to control quality; generally performed after placing orders.

Systems audit. A review of the quality manual and other quality system documents to ensure outside quality requirements are being met.

Third-party audit. See Extrinsic audit

Uptime. Total time that a machine is on-line and capable of producing parts.

Vendor appraisal. Assessment of a potential supplier's capability to control quality; generally performed before placing an order.

References

American National Standards Institute/American Society for Quality Control (ANSI/ASQC). (1994a). *Quality Management and Quality Assurance— Vocabulary*. ANSI/ASQC A8402-1994. ASQC. Milwaukee, WI.

American National Standards Institute/American Society for Quality Control (ANSI/ASQC). (1994b). *Quality Systems—Model for Quality Assurance in Design, Development, Production, Installation and Servicing*. ANSI/ASQC Q9001: 1994. ASQC. Milwaukee, WI.

American Society of Mechanical Engineers. (1994). *Dimensional Tolerancing*. ASME Y14.5M-1994. American Society of Mechanical Engineers. New York.

Blanchard, B. S. (1986). *Logistics Engineering and Management*. 3rd ed. Prentice-Hall. Englewood Cliffs, NJ.

Britz, G., Emerling, D., and Hare, L. (1996, Spring). Statistical Thinking. *Special Publication ASQC Statistics Division*. American Society for Quality Control. Milwaukee, WI.

Chrysler Corporation. (1995). *Product Assurance Planning*. 2nd ed. Platform Quality and Reliability Planning. Chrysler. Auburn Hills, MI.

Chrysler, Ford, and General Motors. (1995a). *Advanced Product Quality Planning and Control Plan*. Reference manual. Distributed by Automotive Industry Action Group. Southfield, MI.

Chrysler, Ford, and General Motors. (1995b). *Failure Mode and Effect Analysis*. Reference manual. 2nd ed. Distributed by Automotive Industry Action Group. Southfield, MI.

Chrysler, Ford, and General Motors. (1995c). *Measurement System Analysis*. Distributed by Automotive Industry Action Group. Southfield, MI.

Chrysler, Ford, and General Motors. (1995d). *Quality System Requirements: QS-9000*. Distributed by Automotive Industry Action Group. Southfield, MI.

Chrysler, Ford, and General Motors. (1995e). *Statistical Process Control*. Distributed by Automotive Industry Action Group. Southfield, MI.

Chrysler, Ford, and General Motors. (1996a). *Quality System Requirements: Tooling and Equipment Supplement*. Distributed by Automotive Industry Action Group. Southfield, MI.

Chrysler, Ford, and General Motors. (1996b). *Tooling and Equipment Quality System Assessment: QSA-TE.* Distributed by Automotive Industry Action Group. Southfield, MI.

Excel Partnership, Incorporated. (1994). *36 Hour Lead Auditor Course.* Excel Partnership, Inc. Sandy Hook, CT.

Ford Motor Company. (1985, September). *Continuing Control and Process Capability Improvement.* Statistical Methods Office. Operations Support Staffs. Ford Motor Company. Dearborn, MI.

Ford Motor Company. (1995). *Advanced Product Quality Planning (APQP) Status Reporting Process.* Ford Motor Company. Dearborn, MI.

Griffith, G. K. (1996). *Statistical Process Control Methods for Long and Short Runs.* 2nd. ed. Quality Press. Milwaukee, WI.

Hillier, F. S. (1969, January). X-Bar and R Chart Control Limits Based on a Small Number of Subgroups. *Journal of Quality Technology,* pp. 17–26.

IBM. (May 1996). Process Control, Capability and Improvement. Quality Institute. IBM Corporation. Thornwood, NY.

The International Automotive Sector Group (IASG). (March 1996). *IASG Sanctioned QS-9000 Interpretations.* Chrysler, Ford, and General Motors. Automotive Task Force. Distributed by the ASQC. Milwaukee, WI.

Kanholm, J. (1995a). *QS 9000 Manual and 40 Procedures.* AQA Company. Los Angeles, CA.

Kanholm, J. (1995b). *QS 9000 Requirements.* AQA Company. Los Angeles, CA.

Kanholm, J. (1995c). *QS 9000 in Your Company.* AQA Company. Los Angeles, CA.

Kolka, J. W., and Scott, G. G. (1992). European Community: Product Liability and Product Safety Directives. CEEM Information Services. Fairfax, VA.

Lamprecht, J. L. (1992). *ISO 9000 Preparing for Registration.* Marcel Dekker, New York.

Lamprecht, J. L. (1993). *Implementing the ISO 9000 Series.* Marcel Dekker. New York.

Omdahl, T. P. (1988). *Reliability, Availability and Maintainability (RAM) Dictionary.* Quality Press. Milwaukee, WI.

Pyzdek, T. (1989). *Pyzdek's Guide to SPC, Volume 1: Fundamentals.* Quality Press. Milwaukee, WI.

Pyzdek, T. (1992). *Pyzdek's Guide to SPC, Volume 2: Applications and Special Topics.* Quality Press. Milwaukee, WI.

Shingo, S. (1986). *Zero Quality Control: Source Inspection and the Poka-Yoke System.* Productivity Press. Cambridge, MA.

Society of Automotive Engineers and National Center for Manufacturing Sciences, Incorporated. (1993). *Reliability and Maintainability Guideline for Manufacturing Machinery and Equipment.* National Center for Manufacturing Sciences, Incorporated. Ann Arbor, MI.

Stamatis, D. H. (1995a). *FMEA: From Theory to Execution.* Quality Press. Milwaukee, WI.

Stamatis, D. H. (1995b). *Understanding ISO and Implementing the Basics to Quality*. Marcel Dekker. New York.

Stamatis, D. H. (1996a). *Documenting and Auditing to ISO 9000 and QS-9000 Requirements*. Irwin Professionals. Burr Ridge, IL.

Stamatis, D. H. (1996b). *Integrating QS-9000 with Your Automotive Quality System*. 2nd ed. Quality Press. Milwaukee, WI.

Taguchi, G. (1987). *System of Experimental Design: Engineering Methods to Optimize Quality and Minimize Costs*. Volumes 1 and 2. Quality Resources (formerly Kraus International). New York.

Index